JUTE AND SUBSTITUTES

VALUE OF EXPORTS OF INDIAN PRODUCTS FOR 1929-30

Rs. 3,108,055,000

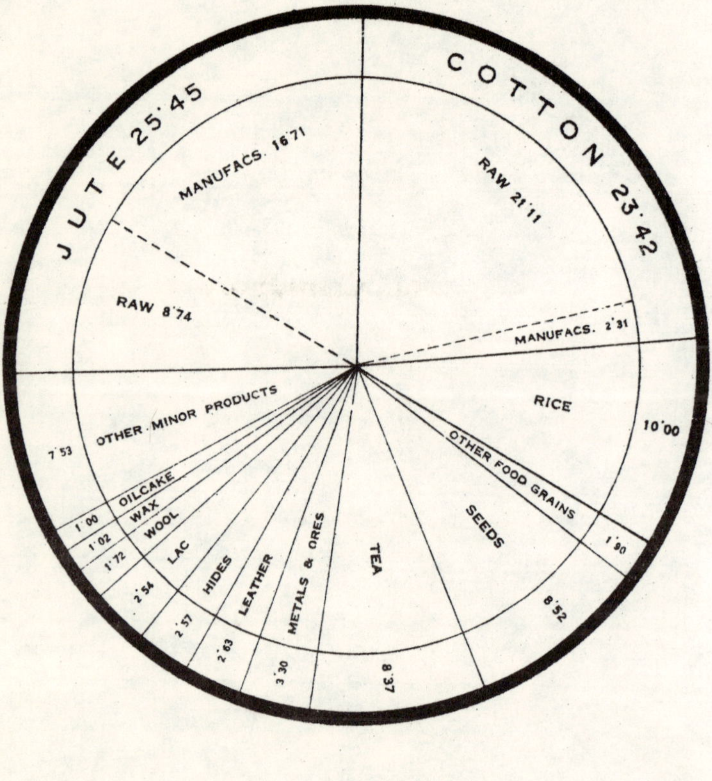

JUTE AND SUBSTITUTES

N.C. Chaudhury

2000

BIOTECH BOOKS

DELHI-110035

Third Edition 1933
Fourth Edition 2000

ISBN 81-7622-049-3

Published by : **BIOTECH BOOKS**
1123/74, Tri Nagar,
Delhi-110035
Phone : 7109765

Printed at : Chawla Offset Printers
Delhi-110052

PREFACE TO THE PRESENT EDITION

THE present edition of " Jute in Bengal " is published under a more comprehensive and appropriate title, " Jute and Substitutes," for the book has not only been thoroughly revised but added to and all possible substitutes for jute described and discussed.

The jute-crop and the industry have been passing through a crisis never known before which has caused general distress throughout the country. It is however hoped that for this very reason the book may be found more useful than before.

My thanks are due to Mr. Sudhindranath Chatterjee, the Librarian of the Calcutta Commercial Library, for affording facilities in my compilation of statistical articles and statements. I must also express my indebtedness to Mr. J. F. Doyle of Messrs. W. Newman & Co., Ltd., for reading the proofs and because of the great interest he has taken in the production.

N. C. CHAUDHURY.

PREFACE TO THE SECOND EDITION

THIS book was first published in 1908. One thousand copies of this edition were sold off within two years from the date of issue. I gratefully appreciate the approbation with which it was received by the public both in India and the United Kingdom. The book has now undergone a thorough revision and some new chapters have been added to it.

Jute is one of the most important crops in Bengal and a very valuable asset of the Indian Empire. But the landless middle and poorer classes of Indian population complain often-times against the cultivation of jute, as they hold that this crop is responsible for the high prices of food grains. I have endeavoured in these pages to convince my readers that high prices have little to do with the cultivation of jute. On the contrary, the growers of, and dealers in, jute have materially been benefited by the extension of its cultivation, and many others have greatly been relieved of the pressure which would otherwise have been more acutely felt by them.

As a touring officer of the Agricultural Department of Bihar and Orissa, formerly of Bengal, I have been closely connected with the crop for about 25 years. So far back as 1898, I started, on my own initiative, an enquiry regarding the races and qualities of jute grown in Backergunge (Chapter IV—Part III). On receipt of my report, which was forwarded to him by the Director of the Department of Agriculture, Bengal, for an expression of opinion, Sir George Watt, then Reporter on Economic Products to the Government of India, was pleased to say "the note contains much of considerable value and interest" (Reporter's letter No. 796, dated the 21st March, 1899, to the Director of Agriculture, Bengal), and he desired that the Department of Agriculture, Bengal, should initiate, in other districts, similar enquiries regarding the races of jute. Thereupon, the Hon'ble Mr. P. C. Lyon, C.S.I., C.I.E., the then Director of Agriculture, Bengal, was pleased to entrust me with this important task. In the meantime, in 1901, Mr. W. Parsons, Secretary to the Calcutta Baled

Jute Association, had addressed the Director of Agriculture, Bengal, on the subject of the alleged deterioration of jute and asked for assistance in connection with the question of improving the qualities of the fibre. As a result, field experiments were undertaken at the Burdwan Farm and I was selected to conduct the experiments. In regard to this work, Dr. D. Prain, M.A., M.D., F.R.S., as Chairman of the Sub-Committee, Board of Scientific Advice, was pleased to report to the Board as follows :—

> " After visiting the Burdwan Farm and carefully examining the sample experimental plots, I am of opinion that the experiment has been carefully thought out and is being carefully conducted." (*Vide* Chapter II—Part III—on Jute Experiments in Bengal.)

The raiyats cultivate jute without any discrimination as to its race. In fact, they sow seeds of different races together on the same plot of ground, causing the act of isolation of a race for a botanical study difficult. During 1899 and 1900, numerous botanical specimens of different races of jute, cultivated in the jute-growing districts in Bengal, were collected by me and forwarded to the Reporter on Economic Products to the Government of India. In my reports on the Jute survey work, special attention was invited to *Kakya Bombai* jute which I discovered in 1899, in a field in a village in the Serajganj Sub-division, Pabna (Chapter V—Part III—on Races of Jute in Pabna). As usual, it was grown indiscriminately with other races. I distinguished this race from others, as it appeared to be the most prominent, and named it as " *Kakya Bombai* "—the name by which the crop of the mixed races was designated by the raiyats of that locality, where it was just then introduced from Goalpara in Assam, by a seed-dealer. Experiments carried out afterwards, under my personal supervision, at the Experimental Stations (*vide* Chapter III—Part I—on Races of Jute) proved it to be one of the best and most promising races. It was mentioned prominently, therefore, in this book published in 1908. Continuous and thorough trials have since been given to this race for a number of years by Expert Officers of the Agricultural Departments, Bengal, Assam and Bihar and Orissa, at various places. It gives me great pleasure to find that it has maintained its superiority and it is now the most favourite of all the races of jute cultivated in the

jute-growing Provinces. In opening the first annual meeting of the Board of the Bengal Agricultural Department, in March, 1920, His Excellency the Governor of Bengal, Lord Ronaldshay, was pleased to remark :

> " *Kakya Bombai* yields, on the average, about 2 maunds of fibre more per acre than the average local races, and the demand for the seed already outstrips the supply. In 1918-19 this selected jute was grown on an area of 100,000 acres, and the increased yield of fibre on this area is estimated to have been 200,000 maunds, worth Rs. 20,00,000."

I gratefully acknowledge the assistance rendered by Mr. D. N. Ghosh, Officiating Director of Statistics, Government of India, and Mr. H. M. Haywood, Secretary to the Bengal Chamber of Commerce, in placing the statistical records of their offices at my disposal.

I shall fail in duty, if I do not acknowledge the debt of gratitude I owe to the late Mr. James Luke of " Capital," to whom the popularity of this work is chiefly due. It was Mr. Luke who brought the first edition of the book prominently to the notice of the trades-people both in India and the United Kingdom.

N. C. CHAUDHURY.

CONTENTS

PART IV—TRADE

PART V—INDUSTRY

PART VI—MISCELLANY

STATISTICS

APPENDICES

INDEX

ILLUSTRATIONS

JUTE
AND SUBSTITUTES

PART I—INTRODUCTORY

CHAPTER I

The Jute Plant

THERE are few people in Bengal who have not seen the jute plant under cultivation. The industry which has grown from it is truly amazing. It is entirely a monopoly crop of Bengal. All attempts to introduce it into other countries, with the exception of China, where it grows to a limited extent, have hitherto failed. Other plants were tried, but none could compete with jute, in points of outturn and cheapness of fibre.

The jute plant thrives on the new alluvial soils formed by the rivers Ganges and Brahmaputra and their tributaries, with rainfalls and temperatures peculiar to the jute growing tracts, downwards from Purnea in Bihar and Orissa and Goalpara in Assam, avoiding the red old alluvial and brackish soils. This area comprises about 5,000 square miles, of which the old Brahmaputra area (about 1,000 square miles), consisting of the eastern parts of the districts

Mymensingh and Dacca and the western parts of the district Tipperah, through which a slow course of the Brahmaputra flows down, is the typical land where a superfine quality of the fibre is obtained.

Two Different Species

Jute is a herbaceous plant of different races and colours, comprising two different cultivated species, one being distinguished by round ragged capsular fruit known in Western Bengal as " Belati Pat," while the other bears a long cylindrical ribbed fruit called " Deshi Pat." In Eastern Bengal the former is called " Deshi Pat " and the latter " Bangi Pat," which is known by the name of " Tosha Pat " in Northern Bengal. The leaves of the capsular species are bitter in taste, while those of the other species are not so; the first is therefore called " Teeta (bitter) Pat " and the second " Mithha (sweet) Pat." The leaves of the latter are largely used as a pot herb and those of the former, after being dried, as a medicine in the rural areas all over Bengal. The most important part of the plant, the fibre which supplies the world with the cheapest sacking cloth, was not, perhaps, known to the ancient Hindus; although it is believed by eminent Western scholars, including Sir George Watt, that poorer classes of Indians were largely clad in jute cloth, even up to the middle of the Nineteenth Century. There is a Sanskrit word " Nalita," probably for the jute plant used in medicines, but there is no mention of its fibre being used in fabrics. We find in Manu's scripture, the names of different cloths woven with different threads, *viz.*, " Karpasham " (cotton), " Sana " (San hemp), " Khoumah " (linseed fibre), and fleece. There is also mention of " Angshu Patta," a fibre of some bark; probably this particular " patta " (cloth) was being manufactured from true hemp fibre of the *ganja* plant (*Cannabis sativa*) which was undoubtedly known to ancient India. It is not proved that this fibre (" Angshu patta ") was produced from the bark of " Nalita " mentioned above. As stated the jute plant does not grow on the high lands of Upper India (Aryabarta—the Aryans' places), the probability is that the fibre was not known to the ancient Aryans, and the plant was grown, to a very small extent, in their kitchen gardens, for a pot herb or a medicinal plant. The fibre was, perhaps, extracted by the Bengal cultivators in the later period when a necessity, for some stuff, easily available, for domestic uses, arose. Evidently

the poorer classes of this country used to put on clothes made with "San" hemp which was cultivated all over India, and it was certainly much stronger and more durable than what could have been made of jute. The Bengalee cultivators who found out the fibre of the jute plant, utilised it in making ropes and gunny bags for their own uses. Afterwards the gunny bags were brought to the local markets and sold off readily, as they were much cheaper than those made of San hemp, or true hemp, which must have been used by the people before the discovery of the jute fibre.

Hand-Loom Industry of Jute

A class of cultivators named "Kapalees" sprang up who specialised in the manufacture of gunny bags by means of hand-looms. Prior to 1857, the exports of jute manufactures from India represented all hand-loom fabrics, the value of these exports amounting to Rs. 21,59,780 in 1850. Since 1854, the hand-loom industry of jute rapidly declined owing to the manufacture of gunny cloths in Europe by mechanical appliances, which was firmly established in Scotland, by this time. During the year 1828, the first consignment of the jute fibre amounting to 364 cwts. (500 maunds) was exported from Bengal. It was, in the beginning, used for the manufacture of cordage and shortly afterwards for that of gunny bags. Thus, this important industry of jute manufacture of Bengal by hand-power was ruined. Up to the year 1853, no effort was made to introduce European machinery into India for the manufacture of jute. The first jute mill was established by George Ackland, a Ceylon Coffee planter, at Ishra in Hooghly, during the year 1854. Since then, factory after factory sprang up all around Calcutta, and there are now not less than 90 mills, each mill employing, on an average, 3,000 labourers.

CHAPTER II

Objections to Jute Cultivation*

AN agitation has, of late, been set on foot, in some quarters, against the cultivation of jute. The following are the three definite charges formulated against the growing of this crop :—

(i) Jute is responsible for famine and the high prices of food grains in Bengal.

(ii) Jute brings about malarious fever in Bengal.

(iii) Jute makes the raiyats extravagant.

We shall try to prove that these charges do not stand on facts.

First Charge

It is true that since 1828, when the exportation of jute to Europe commenced, the cultivation of the crop has expanded in Bengal and Eastern Bengal and Assam with rapidity. But the total area under jute does not, even now, exceed 6 per cent. of the total cultivated area in these two provinces. On the other hand, the area under rice, which is between 60 and 65 per cent., does not appear to have been reduced, to any extent, within the last 20 years or so. The increase of jute area, without reducing the area under rice, may be accounted for by the fact that the area under cultivation has also considerably increased, especially in Northern Bengal. It will also be remembered that jute is immediately followed by transplanted paddy in the Northern districts where jute is very extensively cultivated. Thus two crops are obtained from one and the same land, in the same season, without any appreciable harm to either. But, assuming that jute has reduced the area under rice, we should not, even then, grumble, as it pays the raiyats better. Further, jute is an early crop, harvested at a time when the raiyats are particularly in need of money. Again, in an emergency, jute may be cut at a comparatively early stage, bringing immediate relief, while in the case of the other crops, the raiyat is obliged to wait till they are ripe.

* This article, written in 1907, is reprinted without making any alterations.

There are also other reasons in favour of jute. The paddy crop, at times, fails in Eastern and Northern districts, owing to floods, which generally come in after the jute is off the ground. Thus a raiyat is sure to get at least one crop, even in such a bad year. It is also known to all intelligent raiyats that no crop does well on the same land year after year. Therefore, it is good to follow a system of rotation of paddy with jute where the soil is suitable for both of these crops. Owing to these and other reasons, the area under jute in Bengal was not very small even some 20 years back, when this crop was not considered nearly as paying as now.

The present abnormally high price of rice is chiefly due to two consecutive bad seasons in 1905-06 and 1906-07. The crop was seriously damaged by floods in Eastern Bengal and Assam in the former year and in Bihar in the latter, the net shortage, taking the two provinces together, amounting to 23 and 26 per cent.* respectively in the two years. Exports, too, have been far up of late, though they fell off to some extent in 1906-07. It is not necessary for our present purpose to adjudicate as to what extent bad seasons, and in what measure foreign exports were responsible for the present high prices, but the fact remains that, but for jute, the situation in Eastern Bengal would have been far more serious. On the other hand, the condition of the raiyats of North Bihar would not have been so bad, this year, if they had jute. From my long experience, as an Agricultural Officer of Bengal, I am decidedly of opinion, that the jute crop has, on the whole, improved the condition of the raiyats of the jute-growing districts who form the bulk of the population. It has been estimated that in 1906 jute brought in about $15\frac{1}{2}$ crores of rupees to the cultivators as a clear profit.

Second Charge

There are hardly any facts to prove that jute is responsible for malarious fever in Bengal. If malaria had anything to do with jute, then there is no explanation to offer for its prevalence in other parts of India which do not grow it. On the other hand, malaria is not severe, if at all present, in some parts of Mymensingh, Dacca, Faridpur, Tippera, Rangpur and Bogra, where jute is very extensively cultivated. Goalpara and Sylhet are the only districts in

* In the two provinces of Bengal and Eastern Bengal and Assam the normal outturn of cleaned rice may be estimated at about 52 crores of maunds.

Assam where jute is cultivated to any extent. These districts are almost free from this disease.

Malaria prevails badly in localities where the drainage is imperfect, and where, in the vicinity of dwelling houses, dead rivers and stagnant ditches are found overhung by thick trees and shrubs obstructing the sun and free air. These are the villages which suffer most.

The modern theory of malaria is that it is contracted through the bites of a species of mosquitoes whose favourite breeding grounds are stagnant pools and marshes. From my own long experience of jute cultivation, I can express fully that I did not find any mosquito grubs in any water which is used for jute steeping successively. It is believed that mosquito refrains from laying eggs in this water which is saturated with decomposed jute leaves.

Of course, jute-steeping may cause drinking water to become foul, where reserved tanks or wells are not available. This filthy water may produce diarrhœa, indigestion or some other diseases rather than malaria. In fact the cultivators themselves never attribute any unhealthiness to jute-steeping. The odour is certainly not nice, but the discomfort to the nostrils is not without a very substantial set-off, which is in plain words Bengal's chief wealth.

Third Charge

The third charge brought against jute is that the crop which is a source of ready money has helped the raiyats to become extravagant. This charge hardly deserves any serious notice. After the harvest of jute and the completion of their hard labour, if the raiyats spend small sums of money on some *hilsha* fish, milk or sweetmeats, none should grudge them such small comforts. On the other hand, we ought to rejoice at this. Even if there is any indulgence at present, we may hope that by and by they will of themselves learn economy. It will be a better training for them than the forced economy of poverty. We should be rather proud of our Bengalee Hindu and Muhammadan raiyats, who unlike labourers in other lands, do not spend a single copper on drink.

In conclusion, I may remind my readers that from an agricultural point of view, Nitrogen is the governing factor in the fertility of the soil. Any country, therefore, that exports nitrogenous compounds loses fertility, whilst a country that imports such compounds gains in fertility at the expense of the former. The export of food grains from India not only means the stiffening of present prices,

but, what is far more serious, it means increasingly diminished outturns in years to come. On the other hand, the export of fibres (such as cotton and jute), oils (as distinguished from oil-seeds), sugar and starch which are all non-nitrogenous products, from a country, does not, in any way, impoverish its soil. Add to it the fact, that in our country, we have no special deposits of nitrogenous manures, and, unlike the farmers of other countries, the Indian raiyats would not buy such manures on account of their poverty and ignorance. From the point of view of agricultural economy, the export of jute fibre cannot, therefore, be considered harmful.

CHAPTER III

Chemistry of Jute

CELLULOSE forms the chief constituent in all the vegetable fibres, varying in its purity in different fibres. Cotton is almost purely a celluloid fibre.

Bast fibres such as jute, hemp, flax, etc., contain lignine in a greater or lesser degree. The cellulose of these fibres is called ligno-cellulose which is changed more or less into woody tissue. Cotton consists of one single cell with little incrusting matters, while the bast fibres are made up of an aggregation of cells bound together in a compact form with more or less gummy and resinous matters together with oils, mineral matters and lignified tissue. This lignified tissue lowers the economic value of the bast fibres. Highly lignified fibres are stiff and brittle, and they are little adapted to the spinning of fine yarns. There are more or less pigment matters in all the vegetable fibres, brownish or yellowish colours predominating. The colours are due to incrusting impurities. When fibres are purified or freed from such impurities, they are white and possess a high degree of lustre. Purified fibres are more elastic than those containing impurities: fibres containing more lignified matters are less elastic.

Albuminous matter is found in the dried tissue of the bast fibre. Silicic acid is sometimes found in epidermal cells. Silicic matter in the ash is not dissolved in Hydrochloric acid, whereas the rest is readily dissolved by it. Crystals of calcium oxalate occasionally occur. It is insoluble in acetic acid, but dissolved in Hydrochloric acid. There is also cutose originated from plant wax. It is slightly soluble in caustic potash, but not soluble in Sulphuric acid. It is dissolved by the action of alkalies and hydrochlorides in bleaching. Woody fibre is readily detected by Aniline sulphate reagent, which gives it a golden yellow colour. When treated with Nitric acid, or boiling with caustic potash under pressure, most of the lignine matter is removed, exhibiting characteristics of ordinary cellulose.

Fibres which contain only pure cellulose are coloured blue when treated with Iodine Sulphuric Acid; while fibres

containing lignine are coloured yellow to brown with the same reagent.

This reagent colours cotton blue and jute yellow to brown, and Ambari hemp, San hemp, and flax are coloured blue to copper red. Cotton and San hemp treated with Aniline sulphate solution are not changed, while jute changes to orange colour. Vegetable fibres are not weakened or disintegrated by the action of dilute alkalies. Free Sulphuric or Hydrochloric acid of moderate strength will quickly attack the fibre. Vegetable fibres are more conductive of heat than the animal fibres. Hence animal fibre burns very slowly, while vegetable fibre burns very quickly. Fabrics made of wool and silk are warmer than those made of vegetable fibres.

Jute fibre is relatively weak in comparison with other fibres. But of its fineness and silk-like lustre, it is of special value in spinning. The chief defect is its lack of durability. It rapidly deteriorates when exposed to dampness. Even under ordinary conditions, it gradually becomes brittle and weaker. Bleached fibre is also liable to such deterioration. Hence it is principally used in weaving cheap materials such as sacks, rugs, ropes, twines, etc. San hemp fibre is better than jute in its textile strength, especially when it is liable to exposure.

The comparative strengths of various cordage fibres which follow have been taken from Royle's work on the Fibrous Plants of India. These tests were made in ropes of the same size and of 1·2 metres in length.

	Dry Kilos.	Wet Kilos.
San hemp ...	... 51	72
Jute (*C. capsularis*)	... 65	66
Do. (*C. olitorius*)	... 51	56
Hibiscus cannabinus	... 52	60
(Ambari hemp, Amlapat, Chana or Mestha)		
Rozelle (*Hibiscus Sabdariffa*) ...	... 41	53
Ramie ...	... 110	126

Ramie fibre or Rhea or Kankura of North Bengal is the strongest of all the bast fibres. It is fine, durable and of a lasting silky lustre, which readily absorbs dye.

Chemical analyses of jute and San hemp show that there is about 65 per cent. of cellulose in jute and about 80 per cent. of cellulose in San hemp. But selected fine jute often contains as much as 77 per cent. of cellulose and from fine Mestha fibre about 75 per cent. of cellulose is obtained. It all depends on the cleanliness of the fibres.

	Jute.*	San.
Water (Hygroscopic)	9·93	9·60
Aqueous extract	1·03	2·82
Fat and wax	0·39	0·55
Ash	0·68	0·61
Incrusting and pigment matters	24·41	6·41
Cellulose	64·24	80·0

In connection with the Chemistry of the jute fibre, something may be said about the stalk of the plant, a bye-product, left after the fibre is removed from the stems.

It is not less than a ton and a half per acre which is either wasted in the pools in which the fibre is extracted or used for fuel purposes. The following results of the chemical examination conducted by us will show that the stalk, the bye-product of the jute fibre, contains about 60 per cent. of woody fibre and little ashes. This woody fibre, which may be abundantly available at nominal prices, will probably form a most suitable material for paper manufacture.

Results of chemical examination of jute stalk (bye-product of jute):—

Water (Hygroscopic)	5·85 %
Protein	1·40 ,,
Ash	0·92 ,,
Carbohydrate	31·33 ,,
Fibre	60·50 ,,

The fibre was estimated in the usual way of Sulphuric Acid and Sodium Hydrate treatments.

* The composition of jute may be represented by the empirical formula $C_{12}H_{18}O_9$, *i.e.*, Carbon 47, Hydrogen 6 and Oxygen 47 parts in every 100 parts of the fibre. It contains no Nitrogen.

CHAPTER IV

Races of Jute

JUTE is an annual herbaceous plant of the natural order *Tiliaceæ.* It grows all over Bengal, excepting only the laterite tracts of West Bengal and North Bengal. In Bihar and Orissa the cultivation is confined to Purnea, parts of North Bihar and some parts of Orissa. It is extending in some districts in Assam, and is also grown in the plains of Nepal, towards the Purnea district.

There are two cultivated species of jute, *e.g.*, *Corchorus capsularis* and *Corchorus olitorius*. *C. capsularis* is easily distinguished by its rounded capsules, while *C. olitorius* bears long cylindrical pods. The seeds are very small, of pyramidical shape, with three edges and two apices at both ends, one being conspicuous. The seed of *C. capsularis* is three millimetres in length and two millimetres in breadth and deep brown in colour, while that of *C. olitorius* is two millimetres both in length and breadth, and blue in colour. The races of *C. capsularis* are more numerous than those of *C. olitorius*. A jute plant grows ordinarily to a height between 10 and 12 feet and attains a height of about 16 feet when cultivated carefully with suitable manures.

Jute is commonly called *pat* in Western and Central Bengal, *pat* or *koshta* in Eastern Bengal, and *pata* in the Northern districts of Bengal and Assam. In Orissa it is called *nalita* or *jhouta* and in Bihar *patua*. *Pat, pata* and *patua* are synonymous terms, most probably derived from the Sanskrit word *patta,* a kind of cloth. In Bengali *patta* commonly means a cloth from wild silk, *pat-kapar,* while that from cultivated silk is called *garad*. *Pat* also means in Bengali a leaf-sheath like that of the banana tree or a folding of a cloth or a paper laid one over another. The local names of the different races of each species vary in different localities. There is no common vernacular name for either of the two species. For convenience sake *C. capsularis* may be named *Guti pat* from the rounded shape of the fruit of this species, and *C. olitorius* called *Shuti pat* from the pod-like shape of its fruit. The leaves

of *C. capsularis* taste bitter, while those of *C. olitorius* are not so. Hence in some places, the former is called *teeta* (bitter) *pat* and the latter *mithha* (sweet) *pat*. The leaves of *C. olitorius* (formerly known in Europe by the name of "Jew's mallow") are more largely used as a pot herb than those of *C. capsularis*.

C. olitorius is cultivated principally in the districts of Hoogly and the 24-Parganas where it is known by the name of *Deshi pat*, while in Eastern Bengal it is called *Bangi* and in Pabna *Tosha*. The Northern jute is mostly *C. capsularis*. In Eastern Bengal, *C. olitorius* is cultivated to a very limited extent.

The fibre of *C. olitorius* separates more readily from its stalk than that of *C. capsularis*, hence this species is preferred in some quarters, especially in West Bengal.

C. capsularis, when it is about 5 feet high, can withstand water submersion up to its middle. *C. olitorius* cannot stand water-logging at all, until it is ready for cutting. On the whole, *C. capsularis* is better capable of withstanding adverse climatic conditions, such as drought or excessive rains or high temperature.

In colour and fineness, the fibre of *C. capsularis* is much superior to that of *C. olitorius*, but it is usually somewhat shorter in length. The colour of the fibre of *C. olitorius* is silky yellow, which is very much liked by the sack-weavers in Eastern Bengal. From this peculiar colour, the fibre of this species can easily be distinguished from that of *C. capsularis*, although the colour of the fibre depends chiefly on the cleanness of the water in which it is steeped and washed.

The fibre of *C. olitorius* is more brittle. Roxburgh found, in his comparative tests of the fibres of India, that a "dry line" of *C. capsularis* broke with a weight of 164 lbs. and a "wet line" with the same weight; whereas *C. olitorius* gave way with 113 and 125 lbs. respectively, the wet line gaining 12 lbs. in weight. This fact of the superiority of the fibre of *C. capsularis* over that of *C. olitorius* is well known. Under the same test a "dry" and a "wet" line of *san hemp* broke with 160 lbs. and 209 lbs. respectively. The best fibre of *C. capsularis* is sold at a higher price than the best fibre of *C. olitorius*.

The colour of living plants of *C. capsularis* varies from light green to purple, while that of *C. olitorius* from light green to pink. In some green-stemmed races, the colour of the stem differs from that of its petioles and fruits. It does not appear that the purple or pink colour of the stems of the plants has anything to do, in particular, with the colour of

the fibre. Generally speaking, green-stemmed races of *C. capsularis* are more liked by the jute cultivators, excepting those living in some parts of North Bengal, who prefer purple-stemmed races, which are said to be more prolific than the paler races. But in Eastern Bengal, the green-stemmed races are considered to be heavier yielders. Our experience corroborates the statement of the raiyats of Eastern Bengal. The bark of the green-stemmed races is thicker than that of the purple-stemmed races. The purple-stemmed races are, however, less branchy.

The same race of jute would yield fibres of different qualities if grown in different places, under different conditions. The low land jute is always inferior to the high land jute. The low land jute which is grown in water is always " mossy " at the bottom and harsh. This portion of the fibre is cut off, before it is spun, and is kept apart for batching purposes. The best sorts of jute available at the Serajgunj and Naraingunj markets are the products of high lands. The *Dowrah* jute of Madaripur, which is sold at the lowest price, was almost equal to the best Mymensingh jute, when grown side by side, at the Burdwan Experimental Farm during 1904. The Madaripur quality of jute has greatly improved lately, by more careful cultivation and preparation. The following table shows the experts' valuation report on the samples of jute submitted to the Calcutta Baled Jute Association for appraisement :—

Commercial name of the jute.	Local name of the race.	Value per maund. Rs.	As.
Dowrah ...	(1) Sut pat ...	7	2
	(2) Udhap pat ...	8	2
Naraingunj ...	Dhalsundar ...	7	12
Serajgunj ...	Baran ...	8	2
Northern or Uttarya	(1) Ausa (early) ...	8	5
	(2) Hewti (late) ...	8	6

The cultivators never select the seeds according to the colours of the plants. In the same field and under the same name different races of a species of different colours are grown. The only selection they make is between the early and late varieties, irrespective of colours of the plants, as when washed the fibres of different races, of course of the same species (*C. capsularis* or *C. olitorius*), are identical. In *C. capsularis*, the races as cultivated by the raiyats are mixed indiscriminately, while in *C. olitorius*, the mixture of plants of different races are not quite so bad.

According to the colour of the stems, leaf-stalks and fruits, the races of jute, as separated by the author, may principally be divided as follows :—

(A) Corchorus capsularis

(1) Light green :—Kakya Bombai and Barapat of Serajgunj, Baron or Barapat of Mymensingh, Dhaleswar of Dacca, Hewti (white) of North Bengal.

(2) Light green when young, afterwards purplish :—Amonea of Faridpur, Deswal of Serajgunj.

(3) Purple :—Meghnal or Nalpat of Faridpur, Agniswar of Dacca.

(B) Corchorus olitorius

(1) Light green :—Desi of Hoogly, Bangi or Dewnallya of Dacca, Satnalla, Bangi or Bomi or Bagi of Faridpur.

(2) Pinkish :—Nileta or Tallah of Mymensingh, Tosha of Serajgunj.

(3) Pink :—Desi Lalpat of Hoogly, Nal Bagi of Faridpur.

Both the species have early and late races. The late races generally yield a heavier outturn than the early ones. Both of these races are however useful. The low lands, subject to floods, should always be cultivated with the early races. Those who wish to cultivate jute on a large scale should sow both early and late races, so that the different races may be harvested at different times. It would not be possible to harvest the crop of all the fields at the same time. Under favourable conditions, jute is a very quick-growing plant. It must be particularly noted that the soil should be rich, and that during the early stages of its growth, the crop should receive proper treatment. Otherwise the plants get stunted, flower early and give a poor outturn. The following are the *selected* races of jute showing whether they are early or late :—

Early

C. capsularis :—

Ausa of Mymensingh	(green)
Ditto	(purple)
Bhadya of Rangpur and Jalpaiguri	(green)
Ditto	(purple)

C. olitorius :—

Bangi of Dacca	(green)
Satnalla of Faridpur	(green)
Tosha of Serajgunj	(pinkish)

Corchorus capsularis

(1) (2)

(1) Reduced to one-half.
(2) Full size.

Corchorus olitorius

(1) (2)

(1) Full size.
(2) Reduced to one-half.

Late

C. capsularis:—

Baron or Barapat of Mymensingh	(green)
Kakya Bombai of Serajgunj	(green)
Deswal of Serajgunj	(mixed)
Amonea of Faridpur	(mixed)
Kamarjani of Faridpur	(green)
Hewti of Rangpur	(green)
Ditto	(purple)
Nalpat of Faridpur	(purple)
Kajla of Dacca	(purple)

C. olitorius:—

Desi Lalpat of Hoogly	(pink)
Dewnallya of Dacca	(green)
Halbelati of Tipperah	(green)
Nailta of Mymensingh	(green)

From the following comparative statements of the outturns of the races of jute grown experimentally at the Burdwan Farm, from 1903 to 1906, it will appear that there is hardly any particular variety or race which is the heaviest yielder all along. Dewnallya of Dacca topped the list in the year 1903, but it was placed much lower down, in the following year. So again, the race which gave the highest outturn in the second year, came down in the next year. The Burdwan Farm seed, which was the same as the Deswal of Serajgunj, had occupied the first place during the last two successive years. This is one of the best races we have. It was particularly mentioned in our report on the Jute Experiments at the Burdwan Farm during 1902, the first experiment carried out at Burdwan, at the instance of the Inspector-General of Agriculture in India. The Burdwan Farm seed seems to have improved by careful selection. Kakya Bombai is a tall jute with few branches. It came out as one of the best. The present selected seed has been pronounced to be the best of all the races cultivated in Bengal. When Kakya Bombai (mixed plant of 3 races) was first found by us at a place in Serajgunj, the cultivators, who had just begun to cultivate it, spoke very highly about its yield and other qualities. They considered it to be the best of all the races of jute grown in the Serajgunj Sub-Division. The seed of Kakya Bombai race was originally imported by some seed dealers from Goalpara district in Assam. Any jute seed imported from Assam is still called by the name of Kakya Bombai. They maintained that the seed of this *jat* of jute produced

at Serajgunj was not such a good yielder as that obtained from Assam. Interchange of seeds was highly recommended by the experienced cultivators of Serajgunj.

Variety Experiments

Name of Jute.	Outturn per acre.
1903 — At Burdwan	
(1) Dewnallya (*C. olitorius*) of Dacca	2,661 lbs.
(2) Kajla (*C. capsularis*) ,, ,,	2,581 ,,
1904 — At Burdwan	
(1) Hewti, white (*C. capsularis*), of Rangpur	2,310 lbs.
(2) Kakya Bombai (*C. capsularis*) of Serajgunj	1,935 ,,
At Faridpur	
(1) Barapat (*C. capsularis*) of Mymensingh	2,996 lbs.
(2) Amonea (*C. capsularis*) of Faridpur	2,991 ,,
At Rangpur	
(1) Kamarjani (*C. capsularis*) of Faridpur	2,644 lbs.
(2) Hewti, white (*C. capsularis*), local	2,310 ,,
At Jalpaiguri	
(1) Baron (*C. capsularis*) of Mymensingh	1,110 lbs.
(2) Nalpat (*C. capsularis*) of Faridpur	1,049 ,,
1905 — At Burdwan	
(1) Farm seed (*C. capsularis*)	2,600 lbs.
(2) Nailta (*C. olitorius*) of Mymensingh	2,400 ,,
1906 — At Burdwan	
(1) Farm seed	2,600 lbs.
(2) Halbelati (*C. olitorius*) of Tipperah	2,480 ,,

Recent Experiments, 1924-25 to 1930-31

Trials on some races of jute have been continued by the Bengal Department of Agriculture. From a perusal of the Department's reports, it appears that D. 154, R. 85 and D. 27 of *capsularis* species and Chinsura-green of *olitorius,* have given, on the whole, satisfactory results. Some local races also did well. But none could claim a superiority, either in yield or quality, every where and every time it was grown. It is the known phenomenal characteristic of the plant.

Samples of jute produced, on the Government Farms and on the agencies of some European jute merchants, were sent to the Indian Jute Mills Association and the Dundee Chamber of Commerce, for appraisement (*vide* Annual Report of the Department, for 1928-29). The valuation of the Dundee Chamber of Commerce shows that the sample of R. 85 from Rangpur was "very good," whereas the same grown at Narainganj, during the same season, was declared to be "very harsh and brittle." Similarly D. 154 yielded "harsh quality and lacks strength" on the Rangpur Farm and a "very fair and good spinning quality" produced in Narainganj. Some of the samples of "local Narainganj" and "local Rangpur" were very highly appraised. Representatives of the Indian Jute Mills Association, on examination of various samples, awarded the highest marks 13, to a sample of the local Akhaura (Tippera) and also to a sample of Fanduk variety grown at Serajganj. Samples of the same races cultivated elsewhere were much lower in valuation. Again, the samples of D. 154 or R. 85 of different places secured 1 mark at the lowest and 12 as the highest. In the opinion of the jute merchants, "locality, environment and seasonal conditions have apparently more to do with the quality of the fibre than the seed or race," an authoritative statement which firmly confirms our experience on the subject. An average higher yield of fibre, per acre, from the Government seed must then be due to diligent attention paid for collection and preservation of the seed.

CHAPTER V

Climate

JUTE is a rainy-season crop. Damp heat is the most favourable for its growth. Excessive rain, saturating the soil with moisture, delays both sowing of the seed and the after-treatments.

From one to three inches of rain distributed in a month, during sowing period, may be considered sufficient. Occasional showers of rain, varying from one to two inches, at intervals of about a week, are most beneficial for the growth of the plants.

The temperature on the tracts where jute is grown hardly exceeds, during the growing season, 100° F. and falls below 60° F. Humidity varies from 65 to 91. It is not possible to grow jute anywhere in Bengal during the cold weather.

The table below shows the normal minimum and maximum temperature of the air, the mean humidity and the normal rainfalls recorded at some typical stations in the jute-growing areas :—

Month. 1	Weather Conditions. 2	Calcutta. 3	Mymensingh 4	Dinajpur. 5	Gauhati. 6	Silchar. 7
March ...	Minimum Temperature ...	58	53	49	54	56
	Maximum Temperature...	103	99	104	94	99
	Humidity	80	82	65	75	80
	Rainfall	1	2	1	2	8
April ...	Minimum Temperature ...	70	67	65	68	62
	Maximum Temperature...	103	100	101	91	92
	Humidity	79	81	81	84	87
	Rainfall	2	5	2	6	15
May ...	Minimum Temperature ...	70	71	66	65	66
	Maximum Temperature ...	105	97	108	95	96
	Humidity	81	82	72	81	85
	Rainfall	6	13	7	9	15

Climate

Month. 1	Weather Conditions. 2	Calcutta. 3	Mymensingh 4	Dinajpur. 5	Gauhati. 6	Silchar. 7
June ...	Minimum Temperature ...	75	72	70	73	73
	Maximum Temperature...	96	94	95	95	98
	Humidity	87	90	88	82	88
	Rainfall	11	18	18	15	22
July ...	Minimum Temperature ...	76	75	72	74	74
	Maximum Temperature...	92	89	92	92	92
	Humidity	89	89	91	83	91
	Rainfall	13	16	15	10	19
August...	Minimum Temperature ...	75	76	75	76	73
	Maximum Temperature...	91	92	93	95	98
	Humidity	89	91	91	86	91
	Rainfall	12	15	13	9	20
Sept. ...	Minimum Temperature ...	75	75	75	75	73
	Maximum Temperature...	92	92	93	93	94
	Humidity	90	91	89	83	90
	Rainfall	10	14	12	6	13

The minimum temperature of water in which jute is steeped should be about 80° F.

Rainfall appears to be too heavy in Silchar (Cachar) to be suitable for the jute crop.

CHAPTER VI

Soils

IN Bengal there are old and new alluvial soils which are called *lal mati* or *khiar* and *pali mati* respectively, the same as *bhangar* and *khadar* in Bihar. The new alluvium is generally found near large rivers, especially in their deltas, and is commonly called alluvial soil *par excellence*. The soils of parts of Orissa, South Bihar and Burdwan Division belong to old alluvium. The whole of North Bihar, Eastern Bengal and North Bengal consists of new alluvium, with the exception of the Madhupur Jungle in Dacca and Mymensingh and the *khiar* in North Bengal. Jute grows on new alluvium, but not on the old. A striking example of this fact may be cited:—In Bogra there is a narrow river called *Karotoya*. Over 50 per cent. of the cultivated *pali* lands lying to the east of this river is sown with jute every year, while the cultivated *khiar* lands to the west of the *Karotoya* contain jute not exceeding 5 per cent. None should select this *lal* or *khiar* soil for jute, unless the character of this soil is thoroughly changed by heavy application of cattle manure, or by green manuring, for three or four years successively. *Khiar* land becomes very hard in a drought and prevents spreading of the roots of the jute plant. This seems to be the chief reason why it is so unsuitable for this crop.

Jute grows to perfection on loamy soils. The rain water sinks quickly into loamy soil; hence it is preferred to stiff clay, which can neither absorb nor part with its water as readily as loam or sandy loam does. On clayey soil, it yields a short crop, while a coarse fibre is obtained from sandy lands. Jute fails to grow if it is water-logged.

It is said that jute will grow in the *salt-impregnated soils* of the Sundarbans. We are, however, not prepared to accept this view as true. The author was at Frasergunj in the Sundarbans, on the coast of the Bay of Bengal, for nearly a year during 1907-08, when many experiments on jute were conducted on various parts of the estate. Jute was successfully grown on high lands about 4 feet above the flood level. The soil (surface soil) of this land contained ·20 to ·25 per cent. of common salt. It failed on the soils

containing more salt. But paddy was successful on the soils which had from ·50 to 1·0 per cent. of common salt. It may be noted that a considerable quantity of the salt of the paddy lands must have been washed off during the rainy season when paddy grows. The soils referred to above were collected for analysis during the month of November, after the cessation of the rains. Of course, in some tracts of the Sundarbans, jute has been successfully cultivated, but this is probably due to most of the salt of the land having been removed by drainage before jute was grown. It would be best to proceed cautiously and see whether jute would grow on soils from which the salts have not been so thoroughly removed. It is very well known that when a soil contains soluble salts in quantities above a certain amount, it is unfit for any crops. Even saltpetre, if present in the soil water in a concentrated form, acts as poison for plants. Dr. Voelcker's experiments show that no crops grow on soils (surface soil) which contain ·2 per cent. of soda, ·4 per cent. of common salt, or ·7 per cent. of sodium-sulphate.

Although a chemical test is not a guide to select a site for the cultivation of jute, the following table, showing the proportions of the chemical properties of a typical soil, would be useful for selection of new sites for the crop :—

Insoluble silicates	77·0	per cent.
Soluble ,,	0·25	,,
Ferric oxide	6·00	,,
Alumina	3·60	,,
Lime	2·00	,,
Manganese	0·10	,,
Magnesia	0·10	,,
Potash	0·50	,,
Soda	0·25	,,
Phosphoric acid	0·20	,,
Carbonic acid	2·00	,,
Organic matter and combined water*	8·00	,,
	100·00	,,

Jute grows on high lands, as well as on low lands which are not liable to submersion before the middle of June. Floods cannot do much harm to the plants of some races once they are sufficiently strong, that is, when they will shortly run to flower. But growth, of course, is stopped. Highland jute is always superior to lowland jute, in quality.

* Nitrogen 0·1 per cent.

CHAPTER VII

Injuries

CONTINUED drought is detrimental to the jute crop. The greatest damage is caused by the heavy rain which may fall just after the seed has been sown or when the plants are still young. Stagnant water during this period is still worse. Floods do not harm some races of jute grown in Eastern Bengal when the plants are in flower.

There are several insect pests which cause considerable damage to the plants when they are young. The worst are three kinds of caterpillars. They are known by the following names :—

(1) Indigo Caterpillar (*Laphygma exigua*); in vernacular Gore, Kathri, Horihari, etc.

(2) Jute Semi-Looper (*Cosmophila sabulifera*); in vernacular Ghorapoka, Chhitapoka, etc.

(3) Hairy Caterpillar (*Diacrisia obliqua*); in vernacular Suapoka, Bichah, Bhua, Sambara, etc.

Indigo Caterpillar

The Indigo Caterpillar feeds on the young leaves of plants. In bad cases they defoliate the plants completely, leaving only the bare stems on the ground.

The moth comes out after dark. The female moth lays about 250 eggs in clusters on young leaves; each cluster containing 50 to 200 eggs. The eggs hatch in two or three days, and the small green caterpillars begin to feed on the epidermis of the young leaves. After this, they spread and feed on the leaves, generally in the morning and afternoon, taking shelter on the ground under leaves during the hottest part of the day. If disturbed, they will curl and drop down from the plant. In a fortnight they become full grown, measuring about an inch and pupate in the ground under leaves, coming out as moths in a week.

Indigo Caterpillar

(*After Lefroy*)

Fig. 1.—Egg cluster of indigo caterpillar. Fig. 2.—The caterpillar. Fig. 3.—The moth.

Jute Semi-Looper

Jute Semi-Looper or Ghorapoka (a horse-like caterpillar), a serious jute pest, feeds on the apical buds and top shoots of jute plants. It damages the crop from June to August till the plants are in flowers. The moth remains concealed during the day and comes out at dusk. At night the female moth lays about 150 to 200 eggs singly on the small top leaves. They hatch in about two days, and the little green caterpillars emerge, feeding on young leaves and buds. A full-grown caterpillar, about one and a half inches in length, walks in a semi-looping motion, forming an arch. If disturbed, it jumps off from the plant for hiding and crawls up the plant again afterwards. In 15 days, it becomes full grown and comes down to the ground to pupate. The moth comes out from the pupæ stage in six to nine days. It generally passes the winter as a pupa in the soil.

Hairy Caterpillar

The Hairy Caterpillar feeds on the leaves of jute, Sanhemp and many other field crops. In case of a bad attack, the whole crop is defoliated. Continued rains are favourable for the activity of this insect.

Semi-Looper, Hairy Caterpillar and Jute Apion

(*After Lefroy*)

Fig. 1.—Jute semi-looping caterpillar. Fig. 2.—The moth sitting on a leaf.

Fig. 3.—Egg cluster of Hairy caterpillar. Fig. 4.—Young caterpillars on a leaf. Fig. 5.—Full-grown Hairy caterpillar. Fig. 6. —Moth sitting on a leaf.

Fig. 7.—Jute Apion. Fig. 8.—Its grub. Fig. 9.—Top shoot withering on account of its injury.

The female moth lays from 400 to 1,000 eggs at night on the lower surface of leaves. They hatch in three or four days, and the young caterpillars begin to feed on the lower epidermis of leaves. After a few days, they disperse in different directions and spread over the whole area. In two or three weeks the caterpillars become full grown, measuring one and a half inch. The whole body is covered with hair. The fully grown caterpillar sheds its hair and pupates in the soil inside a rough cocoon of hair. After about ten days it comes out as a moth.

Jute Apion

There is also a small black weevil called Jute Apion. It does not appear to cause any appreciable damage to the crop.

This insect breeds in the stems of jute. The grub is found to make small tunnels in the young stems near the axils of the leaves and the tops droop down.

The two caterpillars on the top of the Figure appear generally in very dry seasons. Continued rains, on the other hand, are favourable to the Hairy Caterpillar. Plants affected by these insects become branchy, and the fibre obtained from them is specky and harsh. Kerosene emulsion,* which may be applied with an Eclair Vaporiser or any fine syringe, is a sure remedy against all these insects. During 1904 the experimental jute crop on about 5 acres of land at Faridpur was very badly attacked by the *chhitapoka*. Kerosene emulsion was applied by us and the crop was saved, while the neighbouring fields suffered a loss by about 50 per cent. The outturn of the experimental fields has already been given in Chapter IV.

Fungoid Diseases

Diplodia corchori of jute, known by the name of "Black Band disease," has been reported from some parts of Eastern Bengal. Fortunately it causes little damage to the crop at present. *Rhizoctonia violacea,* a Wilt disease, and "Chlorosis" in jute are considered to be more injurious. The seed of the affected area should be treated with copper-sulphate solution before sowing; and the affected plants pulled out and burnt, whenever noticed.

* For detailed instruction for preparing kerosene emulsion please see the part Miscellany.

CHAPTER VIII

Extent of Cultivation

THE cultivation of jute had been rapidly extending owing to the high prices obtained for this fibre till it received a sudden blow from the effects of the great European War. Although peace was declared, and the restrictions withdrawn, there were still diminished exports for some time, due to shortage in tonnage. Normal conditions, however, returned by 1922-23. The jute market has again been in a paralysed state owing to the present trade depression and excessive cultivation of the year 1930, and the cultivators are suffering unbearable losses. During the year following, the cultivators had drastically reduced the crop by about 50 per cent. In spite of all these difficulties, new markets are continually being opened out. There is little hope that the cultivation of jute will, in the near future, extend to a great extent.

Under normal conditions, jute is extensively cultivated in the districts of Mymensingh, Dacca, Tippera, Faridpur, Rangpur, Purnea and Pabna. In Mymensingh nearly one-fourth of the total cultivated land is under jute, which yields about one-fourth of the whole production of jute of these provinces. The other six districts account for about 50 per cent. of the total area under jute in Bengal, Bihar and Orissa, and Assam. Thus three-fourths of the total outturn of jute is obtained from these seven districts. The total area under jute, in a normal year, is about 10 per cent. of the total cultivated land of the Province of Bengal in which jute is grown. Taking the whole cultivated area of these provinces of Bengal, Bihar and Orissa, and Assam into consideration, the percentage of area under jute is at 6%*. The total cultivated area in these provinces is about 56 million acres (including about 543,000 acres in Cooch Behar State), of which about 3 million acres are cultivated normally with jute.

---

* It was 5·5 per cent. in 1906.

In Assam, jute now occupies about 3·0 per cent. of the total area cropped, Bihar and Orissa nearly 1·0 per cent. against 2·5 per cent. and 0·9 per cent. respectively during the year 1919.

The following table shows the percentage of area under jute in the net area cropped in the jute-growing provinces in India for the year 1930:—

PROVINCE.	Total area cropped. Acres.	Total area under jute. Acres.	Percentage of the area under jute (approximate).
Bengal (including Cooch Behar)	24,000,000	3,062,000	12·5
Bihar and Orissa ..	25,000,000	238,000	0·9
Assam	5,600,000	192,000	3·0
	54,600,000	3,492,000	

The experimental cultivation of jute in several indigo estates in North Bihar has been successful, though it is not expected that it will take its place among the regular crops in that tract, for there are many other crops to grow easily, during the rainy season. In Assam proper there was very little jute twenty years ago. Goalpara was the only district where jute was cultivated to some extent (28,000 acres) in 1906. The cultivation of jute should be extending in the Assam Valley where the soil and climate are suitable.

PART II—CULTIVATION

CHAPTER I

Preparation of the Soil

LAND for sowing jute is generally prepared just after the first shower of rain during the latter part of February or March. Intelligent cultivators should never lose the opportunity of early rain. If there be no cold weather crop in the field, the ploughing should begin early in the cold season. Deep ploughing is essential for this crop. It has long roots which go down about one foot below the surface of the soil. Soil should be thoroughly pulverised until a fine tilth is obtained. Clods should be broken by means of a *harrow* or *kodali* or a wooden mallet. No treatment should be given to the land when it is too moist. Five to eight ploughings and four or five ladderings will be often found sufficient. Clay soil may require as much as ten ploughings. The corners of the fields which are not properly broken by ploughs should be grubbed by *kodali* before the last ploughing is given. The harrow and grubber would be useful implements for crushing clods and collecting weeds, but these implements are little known in these provinces where jute is grown. After sowing, there should be no ploughing, unless the soil is very light, but simply laddering to cover the seeds, and to press the land in order to bring the moisture up to the surface of the soil, which helps germination of the seed.

CHAPTER II

Manures and Manuring

JUTE does not appear to be an exhaustive crop. Theoretically speaking, the crop itself, that is the fibre, removes no valuable plant foods, such as nitrogen, phosphoric acid and potash. Analysis of the entire jute plant was made by us and the results of the chemical tests have been noted in the chapter on "Chemistry of Jute." The stalk of jute contains 0·03 per cent. of phosphate. The total ash contents amount to less than 1 per cent. It may definitely be said that the soil would not lose much, if the leaves and unnecessary tops of the jute plants be returned to the fields.

Low lands which receive silts every year from floods require hardly any artificial manures. Jute does particularly well if it follows peas or *khesari* grown specially for feeding cattle on the field. Pulses have special capacity for enriching the soil.

Green manuring with *san* is strongly recommended for jute. It is to be sown in November or December and ploughed down early in February. Green manure is exceptionally useful when the character of the soil is required to be changed. By green manuring stiff clay becomes lighter and the light sandy soil becomes retentive of moisture. Organic manures contained in soils retain moisture and ammonia and keep the land warm. Excellent effects of organic manures are especially noticed on light soils.

Experiments at the Burdwan Experimental Farm have shown that cowdung is the best of all manures for jute. Cowdung is followed by castor-oil-cake. Seventy-five maunds of cowdung or six maunds of castor cake per acre, each containing about 30 lbs.* of nitrogen, may be applied when the land is finally prepared. The following statements on the results of the manure experiments at the Burdwan Farm may be read with interest :—

---

* Mr. Mollison, the Inspector-General of Agriculture in India, recommended 30 lbs. of nitrogen per acre for the Burdwan Experimental Farm.

Old Experiments

Manure.	Quantity of Manure per acre. (80 lbs. = 1 maund.)		Outturn of Fibre per acre, Average of 11 years.
	Mds.	Srs.	lbs.
Cowdung* ...	150	0	1,821
Castor cake ...	6	0	1,448
Super-phosphate ...	1	20	} 1,358
Saltpetre	0	30	
Bone meal ...	3	0	1,252
No manure ...	......		648

The experiments at the Burdwan Station conclusively prove the superiority of farm-yard manure (cowdung) to any other. Next comes the castor-oil-cake.

New Experiments

Manure.	Quantity of Nitrogen per acre.*	Outturn per acre, in lbs.		
	lbs.	1904.	1905.	1906.
Cowdung ...	30	1,985	1,805	1,880
Castor cake ...	30	1,635	1,570	1,860
Bone meal and ...	15	} 1,605	1,590	1,600
Saltpetre ...	15			
Bone meal ...	30	1,085	1,580	1,560
No manure ...	......	1,230	1,545	1,560

All the above manures, excepting super and saltpetre, should be applied to the fields during the preparation of the soil. Super may be used with sowing of the seed, and saltpetre as a top-dressing, when the plants are about one foot high. The latter should always be put on the land on which there is a growing crop, otherwise, as it is readily soluble in water, it would be washed off.

From both the old and new experiments, conducted at the Burdwan Experimental Station, it will appear that cowdung and castor cake containing plenty of organic matters are the best for jute. Organic matters are of special value as manures. They add to the soil the power of retaining moisture, a factor essential for the successful cultivation of jute.

* Cowdung contains on the average 0·5 per cent., castor cake 5 per cent., bone meal 3 per cent., and saltpetre 10 per cent. of nitrogen.

Recent Manurial Experiments, 1924-25 to 1930-31.

The results of experiments with chemical manures on jute, carried on by the Department of Agriculture, Bengal, are encouraging. Neutral Ammonium Sulphate has done particularly well on alluvial silted soils in Mymensingh and Pabna, except on land on the river bank. It produced $30\frac{1}{2}$ maunds of jute, per acre, in Mymensingh, against 20 maunds on unmanured plots, and $22\frac{1}{4}$ maunds in Pabna, in place of $17\frac{3}{4}$ maunds, on the check plots. The value of extra fibre, at the cost of Rs. 8 only for $1\frac{1}{2}$ maunds of the manure, applied per acre, is considerable.

The Department has also tested the value of Ammonium Phosphate and Sodium Nitrate with satisfactory results, on red soils, especially when they are applied after liming.

The Department of Agriculture recommends "Adco," a substance, introduced by the famous Rothamstead Farm, in England, for preparing vegetable compost, from waste straw, leaves and other organic refuse, which would be an efficient substitute for Farm Yard manure.

CHAPTER III

Sowing

SOWING season extends from about the middle of February to the end of May. On the low lands, which are liable to be swamped by floods, sowing is finished by the middle of March. When the land is thoroughly prepared, seed is broadcasted and covered by laddering. The seed is sown at the rate of 10 lbs. of *C. capsularis* or 8 lbs. of *C. olitorius* per acre. To ensure an even distribution of the seed, sowing should be carried out crosswise, that is, the field should be gone over by the sower, in the act of sowing, once from north to south, and a second time, from east to west. A harrow or a ladder or a *henga* (log) should follow immediately to cover the seed properly. Germination takes place within three or four days after sowing, if the soil contains sufficient moisture.

CHAPTER IV

Rotation

IN some places, jute is grown on the same land, year after year. This is an exhaustive system. This system should never be resorted to unless sufficient manures are used. Of course, the lands which are submerged during the rains and enriched with silts, do not generally require any manuring. Fibre deteriorates, if jute is cultivated on the same land, every year, without following a system of rotation.

In the principal jute districts where jute is cultivated on high lands, paddy is transplanted immediately after cutting jute. Thus two crops are obtained in the same season. This is undoubtedly a most exhaustive method, if this land is not heavily manured. It may be noted that jute does particularly well if it follows peas or *khesari* which is grown particularly for grazing.

The following rotation schemes are recommended :—

Two-Year Scheme No. 1

First Year

(i) Jute;
(ii) Rape or mustard.

Second Year

(i) Paddy;
(ii) Peas or *khesari*.

Two-Year Scheme No. 2

(If jute is immediately followed by paddy).

First Year

(i) Jute;
(ii) Paddy;
(iii) Peas or *khesari* (grown for feeding cattle).

Second Year

(i) *Aus* paddy;
(ii) Potatoes, or rape, or mustard, or *san*-hemp as a green manure if no other manure is used for the next jute crop.

One-Year Scheme No. 1

(i) Jute;

(ii) Rape or mustard or peas or *khesari* or potatoes.

One-Year Scheme No. 2

(If jute is immediately followed by paddy).

(i) Jute;

(ii) Paddy;

(iii) Peas or *khesari* (grown for feeding cattle).

The reason as to why *rabi* crops are not generally allowed to precede jute, is that in the Eastern districts jute has to be sown early to prevent its being swamped by early floods. Here the *rabi* crops interfere with the preparation of the soil in time. But mustard is soon off the ground. Peas, etc., for fodder also do not interfere with the early preparation of the soil intended for jute.

CHAPTER V

After-Treatment

WHEN the plants are about six inches high, till the plants attain a height of nearly one foot, a *bida* (something like a rake), called *achra* or *nangla* in North Bengal, is drawn over the land, twice or thrice at short intervals, for the purpose of thinning out and loosening of the earth. After this, the crop is twice weeded or hand-hoed, the operation being accompanied by thinning at the same time. The outturn of fibre depends much on this operation of weeding and thinning. Jute fields should be well-cleaned of weeds. Sometimes three weedings or hoeings are given. The plants should be so thinned as to leave a space of about 6 inches between the plants in the case of *C. capsularis,* and 8 inches in the case of *C. olitorius*. *C. capsularis* will bring forth numerous branches if the plants are thinned out over-much. On the other hand, thickly crowded plants will grow too thin to give a good outturn of fibre. Just before the rainy season sets in, the thin and weak plants are finally uprooted. These plants* should be thrown into a manure-pit. No operation should be given when the fields are too wet, that is, when the earth sticks to the implements. It is very essential that every operation is given just in its proper time. It will harm the crop very much, if any operation is delayed even for a few days. Every treatment of the plants should receive the cultivator's utmost care. Well-treated plants will often grow from 10 to 12 feet high—sometimes, reaching 16 feet.

---

* Nowadays fibre is extracted from these rejected plants and sold as *bach pat* (rejected jute).

CHAPTER VI

Cutting

JUTE is cut from the beginning of July to the end of October. It is hardly ever cut before the flowering stage of the plants. It may be cut any time before it is dead ripe. The time of cutting may be divided into three stages as follows :—

First stage—cutting in flower.

Second stage—cutting when fruits set.

Third stage—cutting when fruits fully develop.

Four weeks are taken by the plants to pass through these stages. Sometimes the cutting is delayed by a raiyat for want of sufficient labour or sufficient water for steeping and washing of the fibre. But if the plants are allowed to get dead ripe, the fibre becomes coarse in texture and dirty reddish in colour. The heaviest yield of fibre of good quality is obtained when the plants are cut at the third stage Those who would wish to undertake the cultivation of jute on a large scale, must begin cutting from the first stage, so that they may be able to skilfully distribute the work of harvesting over a long period. It would not be possible for them, for want of labour and other difficulties, to cut the whole crop at the stage which may be the fittest of all.

Plants are cut with a sickle close to the ground. Plants growing in deep water are pulled up. After cutting or pulling, the plants are tied in bundles and steeped in water.

CHAPTER VII

Steeping

JUTE fibre is contained in the bark of the plant. In the natural state, the fibre is associated with a kind of gum which must be first softened by fermentation and then removed by washing. The fermentation takes place when the plants are cut and kept under water. This is called *steeping* or *retting*.

When the crop is grown on high land above inundation level, the bundles are stacked on the field for two days, before they are removed to a ditch for steeping. This is a good practice, as the leaves of the plants may shed and thus be kept on the soil which has produced them. It also reduces the weight of the bundles which are to be carried away a long or short distance. Calculating on the basis that the fibre is 4·5 per cent. of the green weight, 16 maunds of fibre per acre means a crop with a green weight of 355 maunds or 29,000 lbs. The bundles of a stack should be covered with straw or palmyra leaves. If the plants are directly exposed to the sun, the fibre becomes more or less 'specky.' The bundles are then removed to the nearest pool and immersed. In Western Bengal, where deep water is not generally available for steeping, from one to two layers of bundles are bound together with rejected plants and placed in water. Sods of earth are used there for weighting down the stack. This practice is not well adapted as it tends to discolour the fibre. This may partly explain why the *daisee* jute is much inferior in colour to Eastern Bengal *bangi* or *tosha* jute. Logs of wood should always be preferred for weights.

In Eastern Bengal, where jute is cut in water, the steeping begins at once with the leaves on. Several layers of bundles are placed one over another. The second layer covers the first layer completely, leaving only the top 9 inches, and so on. Finally the last layer is covered with weeds. After five or six days the leaves of the plants shed. At this time the whole heap is completely covered with weeds. This system hardly requires any artificial weight—the bundles immerse by their own weight. Sometimes

cultivators erect bamboo posts on either side of the heap to keep it from floating away.

The plants take 10 to 20 days to ret. When the plants are cut in the latter period of the season, it may take even a month for retting—the period depending upon the maturity of the plants, the temperature and other conditions of the water in which they are steeped. The temperature of water should be over 80°F. After a week or so the plants should be examined to ascertain how far the retting has advanced. The examination should be repeated from time to time, till it is found that the fibre separates easily from the stalk. If under-retted, the gum more or less remains and the fibres stick together. On the other hand, over-retting makes the fibre weak and dull.

The water in which jute is steeped has considerable effect on the quality of the fibre. The fibre steeped in clear water, called *kala jal* (black water), is of a white colour, while that steeped in muddy water, called *baga jal* (*baga*=white, truly speaking grey; *jal*=water), is blackish grey. Stagnant *baga* water gives jute a better colour, because it is not so muddy as the running *baga* water. In the case of "black water," whether running or stagnant, the colour of the fibre is not affected, as this water contains little dirty matter. The steeping of jute in running water takes longer to ret than in stagnant water. In running water, the inside bundles of a heap rot earlier than the outside bundles, producing fibre of uneven quality. Thus when the retting of the bundles within is complete, the outside bundles are not yet ready. On the other hand, when the outside bundles are fit for stripping, the inside bundles will be overdone. It is required, therefore, to break the stack and take away the inside bundles when they are ready for stripping, leaving the outside bundles in the steep for two or three days longer. Stagnant water is to be preferred to running water. Deep water is also to be preferred to shallow water in which the lowermost bundles touch the ground, to the detriment of the colour of the fibre.

The crop is so bulky when it is cut that it is not always possible to carry it a long way off, to get clean water for steeping.

CHAPTER VIII

Stripping and Washing

SEPARATING the fibre from the stem must be finished within a couple of days after retting is complete The principal methods of extracting fibre from the stem are as follows :—

First method.—Each plant is stripped separately so that the fibre remains free and without any entanglement. This fibre sells at a higher price. The stripper, generally a woman, does the work in her own home, instead of doing it in the steeping water as in the other processes. She holds the lower part of the stem of each plant with her right hand and deftly pulls the fibre off with her left hand—the fibre passing between the fore and middle fingers of the right hand in which the stem is held. A skilled woman can manipulate as many as three plants at a time, passing three bunches of fibre of three different plants through the three spaces made with the four fingers of her right hand which holds the stems. A woman ordinarily strips about half a maund of dry fibre, working 8 hours a day. But there are skilled women in East Bengal who can strip more than that quantity. The woman is either a member of the raiyat's own family, or one of the family of a friend of his, who charges nothing for the work; but she gets the stalks or piths which are used as fuel.

Dacca, Faridpur and Barisal are the only three districts where jute is separated in this method.

Second method.—When the stems are ready for stripping, the raiyat stands by the heap and takes up a handful of the plants called *muthha,* which are beaten at the bottom by means of a wooden hand and afterwards broken at the distance of one foot from the bottom. The broken pieces of wood are then thrown away. He now takes hold of the separated fibre by both hands, and jerks the stems forward and backward on the surface of the water. After half a dozen jerks, the fibre is cleared off the stalks. A cultivator generally strips 20 seers of dry fibre a day working from 8 A.M. to 2 P.M. A skilled man strips more, sometimes almost double the quantity. But a person unaccustomed to this work, would not be able to extract more than 10 seers of the fibre.

This method is in vogue in almost all the jute districts in Bengal.

Third method.—When retting is complete, the stripper takes hold of a handful of stems with both hands, swings it round his head and dashes it against the water. This is a process which may correctly be called washing rather than stripping. After the stems are washed in this way, they are dried in the sun. At last the piths or stalks are broken at several places and removed from the fibre. Jute is stripped in this method in Orissa. The fibre which is washed by this method deteriorates in its value.

The first process of stripping jute fibre may further be improved by substituting a bamboo or wooden frame for the hand, with more pegs for fingers than a woman commands in one of her hands. This frame may be constructed in the following way:—

Make a bamboo or wooden frame about 24 inches long and 9 inches in width. Attach some 8 or 10 finger-like thick round wooden or bamboo pegs 9 inches long to the frame, at intervals of about 3 inches from one another. This frame, with its pegs up, may be tied to two posts, about 3 feet above the ground, so that a man can work standing.

Now, first, the fibre at the bottom of a plant is to be loosened and put on the frame for work. Each peg should go between the separated portion of the fibre and the pith of a plant; thus when all the pegs are ready, the worker should take hold of the fibre by both hands and drag it out. It would be more advantageous if two men be employed to work together with one frame. It would undoubtedly be found a simpler process than any. Two men would be able to strip about two maunds of dry fibre in a day.

The separated fibre should always be washed in clean water, if available, preferably in running water, as the separated dirt is removed by the current, without affecting the next bundle of fibre.

Washing is generally done by taking a handful of fibre which is dashed right and left in water and sometimes against it. After washing, the fibre is dried in the sun over a bamboo frame for two or three days and then tied into bundles for market.

The best sort of jute is obtained when it is steeped in deep, clean and stagnant water, properly retted, and washed in clean running water; provided the plants are well grown on high lands not liable to be submerged by floods before cutting.

CHAPTER IX

Outturn

THE official estimate of the average normal outturn of jute per acre is 16 maunds. On well cultivated and manured land, an outturn of even 30 maunds per acre may be expected—24 maunds per acre are commonly obtained. Normal average outturn of fibre is estimated per acre at 3·5 bales (400 lbs. each) in Bengal, 3·3 bales in Assam and 3·0 bales in Bihar and Orissa.

The bye-product of the jute-plant is the stalk which is obtained after the fibre has been removed. It is called *pankati* in Bengal. An enormous quantity of this stuff is wasted. It is only partially used as a fuel or as fencing material by the poorer classes and a very small quantity is also utilised in the manufacture of fireworks. An outturn of about 50 maunds of this bye-product may be expected from an acre of land.

CHAPTER X

Seed

VERY little attention is paid by the raiyats to the selection and preservation of jute seed. A raiyat ordinarily preserves neglected plants on the border of a field for the purpose of obtaining seed from them. This is a pity. Vigorous individual plants, which are not branchy, should be set aside for the purpose. Nor does he select the seeds of different races separately. In the same field and under the same name different races of jute are grown and the seeds are similarly harvested together.

An acre of land may yield as much as 6 maunds of seed. After harvesting, seed should be preserved in an earthen jar (*jala*) mixed with some ashes, and carefully covered with an earthen stopper, preventing the access of insects to the seed. The *jala* should be painted inside with coal tar and dried beforehand. Before sowing, 100 seeds from a lot should be tested between two pieces of wet blotting paper. Good seeds will germinate in a couple of days. The seed may be considered good, if the seeds germinate at the rate of not less than 95 per cent. Considering the present value of jute, it would be suicidal to neglect the quality of the seed. The cultivators of East Bengal, where land is inundated, should always indent good seed from elsewhere. On the other hand, the quality of the fibre may deteriorate if the same seed is sown on the same land for a number of years. Interchange of seeds is therefore recommended.

The jute seed is rich in oil. Five seers of seed may yield one seer of oil if extracted in a crude way. The oil is said to be a specific in skin diseases. In the Hill Tippera State oil is obtained in the following manner. The seed is at first pounded and then moistened with water. This paste is put into a vessel with perforated bottom. Another vessel is placed beneath it to collect the oil dropped down. A third vessel or earthen pot with ignited charcoal within is kept over the first vessel containing the paste. The heat is continued until the oil is fully extracted.

CHAPTER XI

Cost of Cultivation and Profit

THE cost of cultivation and preparation of the fibre of jute varies with the conditions of the localities in which it is grown. In North Bihar and Orissa, the wages of a labourer, in and about the year 1920, were not more than 4 annas per diem, while in Bengal the rate of labour, during the jute season, varied from 8 to 12 annas per diem. The following table will show, in detail, the necessary costs as required in the same period, for cultivating an acre of land, in an East Bengal district, where labour is generally dear. Additional cost would be necessary if the fibre is to be tied in drums or bales for despatch to Calcutta.

Item	Rs.	A.
8 ploughings (1 ploughing requires 3 ploughs @ Re. 1 a plough)	24	0
5 seers (10 lbs.) of seed @ 8 annas a seer ...	2	8
2 rakings with *bida*	2	0
First weeding—24 men—at 8 annas per head ...	12	0
Second weeding—12 men ,, ,, ,, ...	6	0
1 thinning—6 men ,, ,, ,, ...	3	0
Cutting and steeping—24 men ,, ,, ,, ...	12	0
Stripping at the rate of Re. 1-8 per maund (16 maunds)	24	0
Collecting fibre and tying—2 men	1	0
Rent	6	0
Total Rs. ...	92	8

against Rs. 58 estimated in 1908.

Additional costs for manuring :—

Item	Rs.	A.
(i) 100 maunds of cowdung @ 1 anna per maund	6	4
Application of manure requiring 4 men ...	2	0
Total Rs. ...	8	4

against Rs. 6-4-0 estimated in 1908.

(ii) When required quantity of cowdung was to be reduced :—

	Rs.	A.
Cowdung 50 maunds	3	2
Application of manure	1	0
Castor cake 3 maunds @ Rs. 4 per maund ...	12	0
Crushing castor cake	1	0
Total Rs. ...	17	2

against Rs. 9-14-0 estimated in 1908.

(iii) When cowdung is not available :—

	Rs.	A.
6 maunds of castor-oil-cake @ Rs. 4 per maund	24	0
Cost for crushing castor cake	2	0
Total Rs. ...	26	0

against Rs. 13-8-0 estimated in 1908.

The cost of cultivation in Bengal amounted to about Rs. 92, that is, at the rate of about Rs. 5-12-0 per maund of fibre produced, against Rs. 3-10-0 estimated in 1908. An additional expenditure of between Rs. 8 and Rs. 26 is required, if the land is manured. If the yield be taken at 16 maunds per acre, when the land is not manured, the value of the outturn may be estimated at Rs. 128 at the rate of Rs. 8 per maund, leaving a net profit of about Rs. 36 per acre, against Rs. 72 in 1908. It may, however, be noted that the raiyats spend little for cultivation, as it is generally carried out by a system of co-operation called *bodla* in East Bengal and *gathia* in North Bengal. A group of cultivators work together in their fields by turns, for which they provide their comrades on the working days with a breakfast on the fields in the morning and a dinner at about 3 P.M. when a day's work comes to an end. It is a misrepresentation to say that the value of thrift is not known to the cultivators. Owing to the present depression in the jute market (1930 and 1931) it is quite out of the question for the *raiyats* to spend more than Rs. 4 for the production of one maund of fibre. There has already been an all round reduction in wages and if the market continues to be dull, the cost must be further reduced to Rs. 3 for a maund.

As jute responds readily to manuring, it is always advisable to incur the extra expenditure for this purpose. It is quite possible to increase the outturn by at least 6 maunds per acre this way, adding even now Rs. 20 to the gross value, with the cost of manuring not exceeding Rs. 12 for an acre of land. Cowdung should be supplemented by a cheaper oil-cake, when the required quantity of the former is not available.

PART III.—RESEARCH WORK

CHAPTER I

Deterioration of Jute

DURING the year 1898, an attempt was made by us to classify the races of jute grown in Bengal. Accordingly when the Author was on tour in Backergunj, fibres and botanical specimens of different races of jute were collected. These specimens, and our descriptions of the races, were forwarded to the Reporter on Economic Products to the Government of India, for examination. The Reporter, in reply, expressed his satisfaction on the work done, and desired that similar survey work might be carried on in all other jute districts. Thereupon we were deputed on survey work during 1899 and 1900, in Faridpur, Pabna, Mymensingh, Dacca, Tippera, Rangpur and Jalpaiguri, the most important jute districts in Bengal. Along with the botanical work, an enquiry was also made with regard to the alleged commercial deterioration in the qualities of jute of different places. The following is a summary of our studies.

The best fibre is obtained on loamy soil. Plants of clayey soil do not ret uniformly. Sandy soils produce coarse fibres.

Finer quality of jute is always produced on the high lands above the inundation level. The plants grown on low lands (*i.e.,* swamps) give forth adventitious roots freely, producing defective fibre. *Kakya Bombai* jute of Serajgunj does not produce much adventitious roots. It is said by the raiyats of Serajgunj that this specific quality of this variety may be lost if the seed be not changed at regular intervals (*i.e.,* two or three years). In fact, the Serajgunj cultivators get fresh supplies of this seed from Goalpara, Assam. There are many seed-dealers who indent the seed from Assam and sell it at Serajgunj, as *Bombai* or *Kakya Bombai.**

---

* For botanical description, *vide* Chapter V of Part III—Races of Jute in Pabna.

Corchorus capsularis is not to be sown thinly. The plants should be 4 to 6 inches apart from one another. This species has a natural tendency to produce more branches than the *Corchorus olitorius.* Irregular branching shortens the length of fibre. Moreover, it is more difficult to extract fibre from the branchy plants. *Corchorus olitorius* is sown thin, that is, a space of 9 inches is allowed around each plant. *Corchorus olitorius* does not produce much branches. It is asserted that this variety also becomes branchy by long cultivation in the same field. There are some varieties, such as *Kakya Bombai, Baran, &c.,* in *Corchorus capsularis,* too, which do not produce much branches. *Corchorus olitorius* does not grow in swamps.

The fibre of *Corchorus olitorius* is coarse, and therefore it is always sold cheaper than the fibre of *Corchorus capsularis.* The raiyats of Eastern Bengal, therefore, do not like to cultivate *Corchorus olitorius.* The fibre of *Corchorus olitorius* is however liked by the Kapalies, the sack-weavers of Bengal. It has a peculiar yellowish colour. It is never as white, glossy, and fine as that of *Corchorus capsularis.*

It is said that the late varieties, which flower very late (some flower in September), give finer and stronger fibre than the early varieties, which flower generally early in July. Raiyats are not however unanimous on this point. With the late varieties, however, the raiyats get sufficient time for harvesting. In order to suppress the flowering period of the plants, it was suggested by the Imperial Institute authorities to nip off the flower-buds of the plants. It is not practicable, nor is it necessary for the late varieties, if lateness has any influence at all on quality.

The time for the cutting of the plants for the production of fine and glossy fibre is considered by the raiyats to be when the plants will shortly run to flower,* and it has been proved to be quite true† by the Imperial Institute authorities both by chemical and practical examinations. In fact, the raiyats of Pabna and Mymensingh begin cutting, sometimes, about a month before the plants run to flower, *i.e.,* after the plants have grown at least for a period of three months. But generally raiyats like to wait some time longer, in order to get a heavier yield of fibre. The later it is cut, the heavier outturn is believed to be obtained by the raiyats. But Mr. T. N. Mukerjee's experiment† at Rahuta shows that the heaviest yield is

* This jute is called *ful pat.*

† *Vide* Agricultural Ledger No. 37 of 1896.

obtainable from the plants cut just after budding. Mr. Mukerjee's results do not corroborate those of ours. The Ledger referred to does not show which variety was experimented with by Mr. Mukerjee. However, the raiyats of Eastern Bengal finish cutting before the fruits are ripe, in order to get the best result.

The influence of steeping in water on the colour of the fibre is great. Clear stagnant water is the best for this purpose. Retting does not proceed uniformly in running water. Muddy water imparts a black colour to the fibre.

It is believed that the blackish colour of jute in some parts of the jute growing districts is due to iron contained in steeping water. With a view to ascertain the quantity of iron, if any, in the steeping water of the Burdwan Farm, samples of water were forwarded in 1902 to Dr. J. W. Leather, Ph.D., Agricultural Chemist to the Government of India, for analysis.

After making an examination of the samples of water, Dr. Leather made the following observations:—

" I have determined very carefully the quantity of iron present in the waters which have been received from Burdwan, and, as the statement shows, some of the samples contain a small quantity of dissolved ferric oxide. None of them contained any ferrous oxide. The amount of iron present is very small. ·1 pt. per 100,000 is equivalent to ·07 grain per gallon. Whether such small quantity can affect the colour of the fibre is a question which I cannot answer.

" I have left out of consideration all iron which may be present in the silt of such waters. All such silt naturally contains iron, but in an insoluble state. It is true that, in the process of steeping, some part of this might conceivably become dissolved and then affect the fibre; but when one bears in mind that most, if not all, of the water used throughout Bengal for this purpose contains silt, it is highly improbable that the iron of such silt affects the fibre.

" Other constituents of the waters have not been determined. There seems to be no evidence that the salts, which are natural to such waters, can affect the fibre. There is no chemical reason for such an assumption that I am aware of, and, secondly, if the steeping water were affecting the fibre, there should be evidence that damage occurs in specific localities. I see no mention of this in the papers which have been forwarded to my office."

Dr. Leather is of opinion that a very small quantity of iron (ferric oxide), ·07 grain per gallon as present in the

water of the Burdwan Farm, can hardly affect the colour of jute. But from our personal experience of the discoloured fibre of North Bengal and other places where jute is steeped in muddy water, we can say that the quantity of iron, although very small, has something to do to discolour the fibre of jute. The jute areas in Bengal are principally inundated annually by the rivers Ganges and Brahmaputra. The water of the Ganges is muddy, while that of the Brahmaputra is clear. The fact is that the jute of the Ganges area flooded over by muddy water is more or less dark, while that of the Brahmaputra area flooded over by clear water is white.

The chief cause of deterioration in length of the jute fibre is due to the continued cultivation of the crop, year after year, on the same field. It has been observed that the best quality of the fibre is obtained from the fields which are newly broken up, or the fields on which jute is rotated with some other crops. But the raiyats cannot generally spare any land for rotation, owing to one's possessing a small area only.

Where it is convenient and available, cowdung is used as a manure for jute. But in many cases there are no means to carry off the bulky manure to distant fields. The quantity of the dung when supplied is generally too small for the purpose, for the stock of cattle of a raiyat is but limited. It is well known to the raiyats that even the seed of the best variety sown on poor soils produces short staple. On the poor soils the plants flower much earlier, to the detriment in the length of fibre. Farm-yard manure will not be available in sufficient quantity, and in many cases, as already reported, it is not practicable to use it. Some chemical manures are therefore necessary. Saltpetre* (nitrate of potash) and oil-cakes may prove most useful. Experiments in each typical locality are necessary to ascertain the suitable manures and their quantities to be applied most economically.

The chief cause of the deterioration of the fibre as regards its strength and colour, is due to the dishonesty of the *farias* (petty dealers), who purchase the fibre at higher prices and sell it at lower rates. They sprinkle water into the fibre and spread it on sand at night, in order to get its weight heavier. In extreme cases the fibre is altogether

* Unfortunately the results of the experiments with jute at the Burdwan Farm show that saltpetre, super or bonemeal are not suitable manures for jute. Next to cowdung, castor-oil-cake has been proved to be the best.

destroyed, and a great loss is sustained by merchants. We have found some jute containing over 30 per cent. of moisture.* This sort of unfair dealing must be stopped, and it lies principally in the hands of the merchants themselves. They should offer a higher price for a good dry fibre, and reject that which is wet.

To sum up my suggestions for the improvement in the quality of jute :—

(1) Rotation.

(2) Judicious manuring.

(3) Interchange of seeds.

(4) Cultivation of selected varieties or races of jute.

(5) Fairness in the trade.

---

* This we mean above the normal moisture, about 10 per cent., contained by the fibre in the atmosphere of the season, which is, of course, very wet.

CHAPTER II

Jute Experiments in Bengal, 1902 to 1905

THE author was in charge of the jute experiments at the Burdwan Farm and other places (Jalpaiguri, Rangpur, and Faridpur) in Bengal conducted from 1902 to 1905. The objects of the experiments were to isolate or select true types of the so-called varieties of jute cultivated in Bengal; secondly, to find out the races most suitable for particular areas; and thirdly, to test the effects of different manures on jute.

In 1901 and 1902 the seeds of jute were collected from the districts by the officers of the Agricultural Department for experimental cultivation at different centres. The seeds were hopelessly mixed with different races, as the raiyats would grow races of jute together, provided they would mature approximately at the same time. During the first two years of the experiments at the Burdwan Farm it was with great difficulty that the individual races were isolated, and the same names were given to the races as were adopted by the author when he collected specimens for Dr. Watt during 1898 to 1900. Among the isolated races there was also the *Kakya Bombai* jute which was hardly known to the public, but it is now recognised to be the best of all the races of *C. capsularis*. This race was discovered by the author in 1899 in a village in the Serajgunj Sub-division in Pabna. It was not a pure race, as a pure race of jute is never cultivated by the cultivators. The author selected a race out of the mixed races of the *Kakya Bombai* plants cultivated in a field which appeared to him to be the most prominent, and botanical specimens were sent to the Reporter on Economic Products to the Government of India. All the subsequent research work on the races of jute was based on the selection of races of jute made by the author on the Burdwan Experimental Station.

The eminent authorities on Botany, Mr. I. H. Burkill, M.A., Reporter on Economic Products to the Government of India, and Dr. D. Prain, M.A., M.D., F.R.S., Superintendent of the Botanic Garden, Sibpur, were pleased to

visit the experiments at the Burdwan Farm on several occasions. Their notes and remarks about the experiments are reproduced below, which will be of interest to readers.

Mr. Burkill on Jute Experiments at Burdwan.

Section I—Corchorus capsularis.—" I was quite convinced in my mind after my visit to the Burdwan Experimental Farm in August 1902, and subsequent study of the specimens preserved in my office, that *Corchorus capsularis* breaks up into no botanical varieties, but that it has many races, perhaps more or less conspicuously wanting in permanency. The very multiplicity of vernacular names for different forms rather indicates this, and the jute merchants have never done more in the direction of distinguishing kinds than use place-names.

" Last year at Burdwan I noticed how freely insects (particularly the wild bees of the genus *Apis*) visit the flowers and fly from plant to plant. This year I confirmed the observation; and I find that from about 7-30 A.M., when jute flowers open, until late in the day, when they wither, they are constantly visited by a variety of bees. The flowers last but the one day.

" Cross and self-pollination are both effected by the visitors. It would be so contrary to the law of nature for a plant to be insusceptible of cross-fertilisation that I have not tried to prove that cross-fertilisation occurs. I did, on the other hand, experiment with regard to self-fertilisation by wrapping up buds in fine linen, and though I only obtained eight fully-ripened capsules out of the 31 flowers which were wrapped up, the possibility of self-fertilisation was proved; I prefer to think that, to the rain, which soaked the wrappings almost as soon as I had put them on, was due the failure to set seed of the other flowers and not to self-sterility.

" The reverse would, however, strengthen the argument which follows. The bees going from flower to flower are consequently doing much to frustrate the endeavours of those who have charge of the Experimental Farm to keep their races pure; thus, for instance, the race *Barapat* has twice been grown between plots of the races *Baran* and *Kakya Bombai* and is exposed to the possibility of fertilisation from both as well as from more remote plots.

" The abundant opportunities for cross-fertilisation, which occur where the different kinds are grown close together, accounts for the variation which was seen before the plots were thinned in the second experimental sowing

made in 1903 at Burdwan. The opportunity which thinning afforded was used by Babu N. C. Chaudhuri, who has been in charge of the experiments, to get rid of plants not true to type, and to that only is due the nearly general uniformity of each plot.

" As he has undoubtedly set his ideals nearer to him than the cultivator does, his plots are more uniform than the crops will be found to be in the country.

" There are hardly any naked-eye features that one can seize on whereby to distinguish the races of jute except the colour of the stems and leaf-stalks, the time of flowering, and the height that the plant attains. The only one of these characters serviceable at thinning time is the first; and on it as a test the weeding of the crops at Burdwan has been done. Consequently it follows that the race *Barapat* grown between a race with red petioles and another, having, like it, green petioles but beginning to flower a little later, will have weeded out from it the mongrels showing a strain of the red-petioled *Kakya Bombai,* but not those which are the offspring of it and *Baran.*

" It is to be expected, therefore, that if on the closely approximated experimental plots at Burdwan such an experiment as the present one be prolonged for several years, the individuality of the races will be gradually reduced. There will, however, be a tendency to break them up into two classes, *viz.,* red-stemmed and green-stemmed, according to the selection practised in thinning the plots.

" Do the raiyats actually use the colours red and green in any way to weed out undesirable plants? I note that the Rangpur races are particularly red-stemmed and that the Faridpur races all have red petioles on green stems.

" In the country, plots are larger and less approximated; the liability to the production of mongrels is therefore much diminished. It will be at once asked how the races come to be retained. I believe chiefly because each district grows early, mid, and late flowering races, and it seems to me that such an idea should form a good working hypothesis upon which to gather information. A common period is, however, not kept by the races in different districts, but the tendency is evidently for the plant in the northern districts to become quicker-growing than in the southern districts. Facts are wanting which might show any connection between the general lateness of the local races of jutes and the degree of submersion to which the

district is liable. The Burdwan experiments have shown that the Jalpaiguri and Jamalpur races are particularly early-flowering.

"The experiment suggested by this observation is that the behaviour and value of the best Jalpaiguri races should be tested in the neighbourhood of Serajgunj, or in Tippera, Faridpur, and Dacca, which are districts sending the latest kinds to Burdwan; and the reverse should also be tried. It may perhaps be found that imported seed is an acquisition in one or the other place.

"I observed at Burdwan that two plots of plants produced flowers paler than usual. These two were '*Tosha* red'* from Serajgunj (why named red, I do not understand) and *Fulleswari* from Tippera. The only difference between the two, as far as I have observed, is that *Fulleswari* was found to be a little later a little shorter at the dates of my visits. The Burdwan Farm received another kind of jute from Tippera named *Dewdholi*. *Dewdholi* and *Fulleswari* were the only two Tippera jutes. My notes tell me that *Dewdholi* had exactly the same relation to *Kakya Bombai* from Serajgunj that *Tosha* red had to *Fulleswari*.

"It is by no means improbable that seed of *Kakya Bombai* and *Tosha* red from Serajgunj and sown in Tippera would in a few generations become *Dewdholi* and *Fulleswari;* brought back again, in a few more generations they would return to *Kakya Bombai* and *Tosha*. The Burdwan Farm received under the name of *Parbatea* seed from two sources, *viz.*, Jamalpur and Dacca. The Jamalpur plot flowered a little earlier than the Dacca plot; in other respects they were the same.

"If I am right in thinking that the locality does, in a few generations, tend to delay or hasten the growth as suggested above, then this influence will act together with the cross-fertilisation in levelling the Burdwan crops, if they should be grown through a fairly long series of years.

"I unhesitatingly state that—

'*Tosha* red' (Serajgunj) = *Fulleswari* (Tippera) made a little later.

Kakya Bombai (Serajgunj) = *Dewdholi* (Tippera) made a little later.

Parbatea (Jamalpur) = *Parbatea* (Dacca) made a little later.

Barapat (Jagannathganj) = *Barapat* (Serajgunj).

---

* Evidently a mistake occurred in attaching labels. It was undoubtedly "*Deswal* white." *Tosha* is a famous olitorius jute and not capsularis.—AUTHOR.

" I suspect that without difficulty other cases could be established; but to do so would mean an intimate study of the plant in the field. Information in this direction could be collected by the Travelling Inspector of the Department of Land Records and Agriculture, Bengal.

" The Jute valuation made last year by Messrs. Millar, Duffus, and Dott of the samples from the small experimental plots told us only one thing really, *viz.*, that of all the races grown together none was distinctly better or worse than the average of the lot; while of the plots grown elsewhere on the farm, the jute fibre from many was better than from the selected races. This year instead of 11 small experimental plots there have been 36. I am very anxious to know if there is to be the same uniformity in the value of the produce that there was last year. What will be most instructive perhaps from out of the whole series will be the value of the produce of plot No. 16 ' Farm seed.'

" Much depends upon that, and I could wish that a plot of ' Farm seed ' had been sown, say, between every five of the long series, the ground given to which seemed to me to be poorer at one end (towards No. 45) than at the other end."

Section 2—Corchorus olitorius.—" I tried the same experiments with *Corchorus olitorius* as with *Corchorus capsularis* and obtained the same result. I had 28 capsules wrapped in linen and obtained 9 pods. The seeds from these pods sown on a damp tile in my office are now germinating very freely. The flowers of *C. olitorius* are larger than those of *C. capsularis* and are visited by the same insects. Their duration is one day, and pollen is shed around an apparently mature stigma at the moment of the opening of the flower. There is, therefore, the same chance of cross and self-fertilisation in this species as in the other. However, the races are much fewer. I classify them—

No. 1, early green-stemmed, called *Satnala* in the Faridpur district and *Bangi* in Dacca;

No. 2, late green-stemmed, which came as *Nailta* from Jagannathganj, *Deswal** white from Serajgunj and *Paknallya* from Jamalpur;

No. 3, very late green-stemmed, which came as *Dewnallya* from Dacca;

* *Deswal* is a race of *capsularis* jute of Serajgunj. By a mistake, it was labelled as *Deswal* for *Tosha*, already noticed in the footnote of page 56.—AUTHOR.

No. 4, late red-stemmed, received as *Tosha* from Serajgunj and as *Desi lalpat* from Hoogly; also probably here is to be added *Halbelati* from Tippera.

" It is remarkable with regard to the valuation of fibre made last year by Messrs. Millar, Duffus and Dott that they remark of both *Tosha* and *Desi lalpat* that they had stood too long in the field—a remark that they do not make of the early *Satnala* and *Bangi*. They also remark of both the latter that they had soft ends—remarks which confirm my classification by appearance of the plant.

" Altogether the problem of improving the races of *C. olitorius* is infinitely less complex than that of *C. capsularis*."

Dr. D. Prain on Jute Experiments at Burdwan

" After visiting the Burdwan Farm and carefully examining the sample experimental plots, I am of opinion that the experiment has been carefully thought out and is being carefully conducted. It is too soon to say yet this season what the full deductions from the experiments are to be. My visit, which was paid at the request of Mr. Maddox and not primarily in connection with the present reference from the Government of India, was paid in company with Mr. Burkill, who was, I understand, independently under an obligation to pay a visit to the experimental farm. I hope to pay another visit, and further hope that I may be able to pay it along with Mr. Burkill.

" As regards the experiments themselves, I have but one minor criticism to offer. A point to bring out clearly is the extent to which the plants in these plots are subject to cross-fertilisation owing to their being grown in proximity, and consequently to what extent the plots of a second year show, as compared with those of the year before, an increased proportion of plants of mixed type. I should expect this to be considerable. On examining the actual second-season plots at Burdwan, it struck me as wonderfully small. I was, however, informed that when the plants were small and the various plots were being weeded, plants of obviously mixed type were, as far as possible, eliminated. So far as one phase of the experiment is concerned, I realize that this was essential. But I should have liked to see, and had I been consulted earlier, would have recommended, that a small section, say a piece as long as each plot is wide, be left rigorously alone—

weeded to be sure, but with all jute plants of whatever type it might contain, left to grow. The plants would probably have been too thick to serve any other useful purpose, but they would have served to indicate the extent to which the crossing of races takes place. So far I can find no sign that there are any varieties either in *Corchorus capsularis* or *Corchorus olitorius,* though there is apparently a considerable tendency to racial modification, especially in the former species.

" As to the general question of deterioration, it is not clear that this occurs. It is asserted: there is no attempt at proof. I have a vivid recollection of Experimental Jute-growing on the Sibpur Government Farm some ten years ago, and the similarity between the plants seen by me at Burdwan this year and at Sibpur then is most striking. If there be anything in the complaint, the whole tendency of the statements before us is to show that the deterioration lies in more careless retting, not in a demoralised plant. If so, the remedy should be with the trade.

" The report submitted by the fibre experts is the most disconcerting document of the whole series to my mind. It has evidently been drawn up with the greatest care, yet it proves nothing and affords little real assistance to Government. There are several possible explanations of the results of this report, none of them entirely satisfactory to my mind. It seems unnecessary at this stage to discuss them exhaustively. I will, however, indicate the two that appear to me to be most plausible, yet these two being more or less alternatives, do not help us much—(1) It may be safely assumed that the whole of the plots were last year grown, harvested, and retted with equal care. It is certain that the plants in all the plots grew under quite comparable conditions. Yet the results, which may be termed good as regards Deswal jute raised from farm seed, were only indifferent to fair with the most of the other kinds. The suggestion, then, is that seed of a strain natural to, or naturalized in, a particular district gives better results in that district than any freshly imported seed. If this be the case, the experiments at Burdwan, however prolonged, will not lead to the results that Government and the jute industry wish to obtain. They will have to be repeated in every *likely* jute-producing tract.

" (2) It may, however, be supposed that the jutes from other districts grown at Burdwan were really quite as good as these same strains are capable of being in their own districts. It may even be anticipated that in some

cases the change of venue really benefited them. If this is the case, it is not impossible that those in charge of the retting and extraction of fibre at Burdwan being most familiar with the jute that had become so to speak naturalized at the Burdwan farm, get better results from it than was possible from *identical* treatment of the other strains of jute. If this possibility could be proved, the remedy would be to get men from the districts where these particular jutes are naturally grown to go to Burdwan and treat, as they annually treat them, the jutes they are accustomed to. This particular method of overcoming a difficulty is known to and practised on a large scale in certain places by the native raiyats themselves. The following instance, which is a familiar one, illustrates the point. In the beginning of the nineteenth century hemp was grown in the Jessore district for the purpose of producing *round ganja* for the Calcutta market. In North Bengal, hemp was only grown for the purpose of producing *flat ganja*. When the Board of Revenue, Lower Provinces, concentrated the cultivation of hemp for the purpose of producing *ganja* in the present *ganja* mahals round Nowgong and in the adjacent parts of Rajshahi, Dinajpur, and Bogra, obviously *round ganja* for the Calcutta market could only be manufactured in the restricted area. But the people about Nowgong did not know how to make *round ganja* for the Calcutta market: they therefore introduced at the harvesting season Jessore coolies to make it for them. Now though three-quarters of a century have elapsed since any hemp was grown for *ganja* making in Jessore, the Nowgong people to this day import Jessore coolies to make their *round ganja* for the Calcutta market, making the *flat* themselves as they have always done.

"What the raiyats of a particular area can do every year for the special purpose, it is possible that Government might do for one or two years for a parallel purpose—the preparation of a fibre such as is known to suit the Calcutta jute-buyer."

The Results of the Experiments

The objects of the experiments at the Burdwan Farm were to answer the following questions:—

I. Whether the qualities of the fibres of different races differ.

II. At what stage of growth of plants, a crop should be harvested to give the best fibre and heaviest outturn.

III. Whether a well grown crop resulting from suitable manuring would give a better fibre than an inferior crop.

IV. To what extent the quality and outturn of fibre are affected by different degrees of spacing.

V. Whether the retting process can be improved under conditions as ordinarily prevail.

I.—The results of the experiments proved that the qualities of the fibres of different races depend chiefly on the soils and other conditions under which they are grown. The fibres of the same race of jute grown on different fields of the same farm were valued at different prices, while values of the known inferior races of jute were similar to those of the known superior races when cultivated at the same place. The same race of jute yields different qualities of fibres when grown on low lands against high lands, taking that the retting and washing is done rightly in the same water. Jute is a very quick growing plant. If the plants of a race of different plots at the same place are not weeded and thinned out timely, there would be differences in the fibres.

The following table gives the results of valuation made during 1903 and 1904 by some jute merchants* of reputation nominated by the Bengal Chamber of Commerce. The plots at the Burdwan Farm were sown with the *same seed* at the same time but harvested on different dates at the same stage of maturity of the plants, that is, as the pods set after flowering. The crops of different plots appear to be quite different both in yield and value.

Plot No.	Date of sowing.	Date of harvesting.	Outturn of fibre per acre.	Price per maund.
			lbs.	Rs. As.
10	May 18	September 12	1,845	8 12
13	,,	,, 3	1,665	8 3
20	,,	,, 3	1,815	8 12
30	,,	August 18	1,230	8 10
40	,,	,, 29	1,500	8 1
47	,,	September 27	1,680	8 0

The above plots were manured with 5 tons of cowdung per acre.

At Faridpur, on the other hand, all the plots of the same race were sown and harvested at the same time. No fertilisers were applied to the plots during the year under reference, though they were manured with bonemeal similarly during the preceding year. The table below

* Messrs. D. L. Millar, J. C. Duffus and A. S. Dott.

shows that there were much differences in the yields of the different plots :—

		Outturn of fibre per acre.
Plot No.	3	2,610 lbs.
"	12	2,520 "
"	26	2,025 "
"	30	1,755 "
"	34	1,665 "
"	38	1,710 "

Both at Burdwan and Faridpur the weeding and thinning operations were in some cases delayed owing to rainfalls.

The following valuation statement of the fibres of the supposed superior and inferior races of jute cultivated at the Burdwan Farm under the same conditions, proves that there are commercially little differences in the different races of jute grown in different parts of these Provinces. The qualities of the fibres differ owing to their cultivation in different localities, in different soils and lands and produced under different conditions.

Name of the race.	Commercial Name.	Value per maund.	
		Rs.	As.
1. Udhap pat	Dowrah, quality of commerce, Inferior.	8	2
2. Dhal Sundar	Naraingunj, Superior quality in commerce.	7	12
3. Baran	Serajgunj, commercially Superior.	8	2
4. Ausa	Northern, quality Inferior to No. 2.	8	5

In spite of the facts recorded above we are not prepared to minimise the superiority in some respects of some races such as *Kakya Bombai, Hewti, Barapat, Baran, Amania, Kajla, Kamarjani, etc.,* of the species *C. capsularis,* and *Dewnallya, Nailta, Halbelati, Tosha, Desi, etc.,* of *C. olitorius.* The races mentioned above have been found to be heavy yielders (*vide* Chapters on the Races of Jute). There are some other races which are specially suitable for particular lands, as for example the *Sutpat* of Faridpur which can admirably thrive on the lands under deep water where other races would be considered useless. *Kakya Bombai* jute is good on the high lands as well as on the lower lands.

II.—The time of cutting was divided into four periods, *viz.*:—

1st Stage.—Plants were cut when in flower.

2nd Stage.—Cutting when fruit set in.

3rd Stage.—Cutting when fruit fully developed.

4th Stage.—Cutting when the fruit was dead ripe.

The results of the experiments at the Burdwan Farm are noted below:—

1st Stage	...	1,312 lbs. of fibre per acre.
2nd „	...	1,375 lbs. „ „
3rd „	...	1,671 lbs. „ „
4th „	...	1,694 lbs. „ „

The value of the fibre harvested at the fourth stage was less than that offered for the fibres harvested earlier. It was also more difficult to extract fibre of the plants harvested at the fourth stage. Considering all points, the experiments proved conclusively that the best time for harvesting of jute is the third stage, that is, the cutting when the fruits fully develop, but are not yet matured. There is an interval of about 10 days between succeeding stages.

III.—It will be seen that the best grown plants give the best quality of fibre irrespective of differences in characters of different races. Of course it is due only to the length of the fibre. The outturn and price go hand in hand. The following table gives the results of the plots manured with different manures applied in quantity on analysis to supply 30 lbs. of nitrogen per acre:—

Manure applied per acre.		Outturn per acre.	Cost per acre.			Gross value of outturn.	Profit per acre.	Rate per maund at Calcutta.
			For manure.	For cultivation.	Total.			
	Mds.	lbs.	Rs.	Rs.	Rs.	Rs.	Rs.	Rs. As.
Cowdung ...	75	1,985	4	78	82	214	132	8 10
Castor-cake ...	7	1,635	15	74	89	171	82	8 6
Bonemeal ...	10½	1,085	21	74	95	113	17	8 5
Ditto ... / Saltpetre ...	5¼ / 4¼ }	1,605	32	74	106	167	61	8 5
No manure	...	1,230	...	72	72	129	57	8 6

IV.—To what extent the quality and outturn of fibre is affected by different degrees of spacing? To this question, although it is difficult to give any definite answer, the Burdwan experiments proved that a spacing between 4" and 6", in case of *C. capsularis,* is most economical. The crop was grown in four plots, and an attempt was made to keep the plants 4", 6", 8" and 10" apart from each other in the four plots, respectively. The 8" and 10" plots were particularly poor and the fibres were inferior.

Degree of spacing.	Outturn of fibre per acre.	Price per maund.	Date of cutting.	Date of washing.
	lbs.	Rs. A.		
4"	2,079	9 4	September 15	October 3
6"	1,476	8 8	Sept. 16 & 17	„ 9
8"	882	7 14	„ 19 & 20	„ 12
10"	1,183	7 14	„ 20 & 21	„ 13

Taking the first two plots only, which alone were fairly comparable, an attempt was made in 1903 to ascertain approximately the actual spacing in these two plots, as also the number of plants and their green weight at the time of harvesting. The following table gives the result :—

Supposed spacing.	Room allowed per plant (*i.e.* average spacing).		Number of plants per acre.		Outturn of fibre per acre.	
	1903.	1904.	1903.	1904.	1903.	1904
	Sq. in.	Sq. in.			lbs.	lbs.
4"	43	39	150,000	160,000	1,590	2,079
6"	65	63	160,000	92,000	1,050	1,476

It will be seen in both the years that the actual spacing was more like 6" and 8" than 4" and 6", respectively.

V.—Jute is such a bulky and heavy crop that it is difficult to carry the bundles of green plants any distance, and the raiyats are obliged to use whatever water they find

handy. Calculating on the basis that the fibre is 4·5 to 5·5 per cent. of the green weight, we see that 15 maunds of fibre per acre means a crop with a green weight of something like 300 to 350 maunds. The conditions at Burdwan did not admit of retting the plants in running water. The steeping process was similar to that of Eastern Bengal. We did not approve of the local system of covering the heaps of jute plants under retting with sods of earth which certainly discolours the fibre to a certain extent. It is required to give weight to the heaps for sinking. In Eastern Bengal several layers of bundles of jute are placed one over another, which does not require any artificial weight, while in Hoogly and its neighbouring districts, only two layers of bundles are heaped together, as the water available for steeping is generally shallow.

The harvesting operation extended over three months from early August to the end of October.

The temperature of heaps during the retting process was taken; 80° F. was the minimum, which gradually rose to 95° F., the maximum. The temperature fell considerably (70° minimum, 83° maximum) towards the end of October, when the heaps took about ten days more for retting. During the rains the temperature in all the bundles was constant.

The steeping water of the farm was not favourable to get a first class fibre, as is obtained in Mymensingh. The water contained a negligible quantity of iron as shown by the Agricultural Chemist, but other colouring matters from decomposition of vegetable substances were also present.

The time required for retting varied from 15 to 30 days. The smaller heaps generally took a longer time in retting than the larger ones, although steeped on the same day.

CHAPTER III

Fraudulent Watering of Jute

FRAUDULENT watering of jute is in practice more or less in almost all the jute growing districts. This notorious practice is well known to all buyers.

It is done by the unscrupulous dealers known by the name of *bepari* or *faria*. The buyers have to depend on them for the supply. The buyers can hardly check them, especially when the market is on the rise. The dull market prevailing during the last five years has had a considerably good effect in checking this malpractice. Hence the qualities of jute of the markets of the Madaripur, Serajgunj and other districts, which were notorious for fraudulent watering, have greatly improved. The common way of watering jute is to spread out the loose jute purchased by the *farias* on the banks of a river and sprinkle water, sometimes containing sand, on each layer as it is spread out. It is then redrummed and taken to the markets for sale. A few drums of dry jute are always put in front for inspection of the intending buyers.

There is an interesting report on this subject written by Mr. D. N. Mukerjee, M.A., M.R.A.C., formerly of the Bengal Agricultural Department. The following is a summing up of Mr. Mukerjee's report :—

(1) The fine particles of sand stick to the fibre so long as it continues wet, but can be easily shaken off when it gets dry. But the watering is a very much more serious thing. If wet jute is baled or stacked for a few days, the fibre loses its lustre and becomes discoloured and weak—sometimes so weak and brittle that it may be reduced to a fine *powder* by merely pressing it between the thumb and the finger.

(2) Some firms dry the jute before baling, but some others are perfectly reckless about what they bale. Large firms have arrangements for the examination of every drum that is purchased and for drying it properly if found wet before baling. The firms have to keep up quite an army of assorters for this purpose. It is however practically impossible to properly supervise the work of the assorters.

They are men on a small pay, and are constantly passing wet jute as dry on receiving a small bribe from the *beparis*. Thus in spite of strenuous efforts, several of the most respectable firms are powerless to prevent entirely the baling of wet jute. There are other firms that make no bones of the superfluous moisture. They send the jute over to their godowns in Calcutta, where it is sold to the mills or to the shippers. The examination of jute by the buyers in Calcutta is not so thorough as at Serajgunj.

(3) Between the raiyats at one end and the home market at the other, there are the following middlemen: *beparis* or dealers, *mahajans* or *aratdars,* buyers, balers, and shippers. Sometimes the last three functions are combined by the same firm that bale and ship off at Calcutta what their agents buy in the country. But the business at Serajgunj may be said to be practically in the hands of the *mahajans* or brokers. They are well-to-do people, being mostly Marwaris. They advance money to the *beparis* or sellers, the condition being that the latter must bring to the *mahajans* all the jute they can get from the raiyats. The money is not realised from season to season, but is allowed to be in the hands of the *beparis,* one *bepari* sometimes having an advance of Rs. 5,000 or Rs. 10,000, and occasionally no less than Rs. 20,000. When the jute is brought to the *mahajan,* he settles the price with the buyer secretly, the *bepari* having nothing to do with it and often not knowing at all what his jute sells for. He has only a general idea of the state of the market from the price other *beparis* receive for their jute. The *mahajan* gets a brokerage of 2 to 4 annas per maund, but over and above this we may be sure he leaves a margin between what he pays the *bepari* and what he himself gets from the buyers. Plainly he is entitled to some discount for the ready cash he pays in advance, but the transaction might be done openly. The price of jute is subject to violent fluctuations. Probably there is no other article the price of which varies so much in the course of a few days in the same season. Both the buyers and the *mahajans* have daily wires from all the important jute marts telling them of the state of the market, but the information is carefully kept to themselves. It might be an advantage to have some agency for making it public. At some places there are no *mahajans,* but the *beparis* deal directly with the buyers. The raiyats also to some extent sell their own jute to the purchasing firms without the intervention even of the *beparis*.

(4) Even what is generally held in the trade as dry jute is not as dry as it might be. It is not "bone dry" as it is called. If such jute is exposed to the sun, it quickly loses 8 to 10 per cent. of its weight. This amount (which of course represents moisture) does not affect the quality of the fibre in the bale or in the stack. The mills also accept this amount of moisture (10 per cent. up to November and December and 8 per cent. later), but any quantity above this is held as objectionable. The mills now generally insert a clause in their contracts to the effect that deductions will be made for moisture in excess of the aforesaid quantities. It is not at all uncommon to deduct as much as 12 per cent. on the weight delivered of Madaripur jute. Some of the mills have their own agents who would not object to buy wet jute, as it need not be baled or stacked, but used in the mills as soon as it arrives.

(5) Mr. Mukerjee obtained some samples of well-dried jute from different markets which were redried and reweighed by him.

In each case 2 lbs. weight of the fibre was dried in the sun for two days and then reweighed with the following results :—

Sample.	Original weight. lbs.	oz.	dr.	Weight after two days' drying in the sun. lbs.	oz.	dr.	Percentage of loss on drying.
(1) Fine Mymensingh	2	0	0	1	12	7	11·13
(2) Fine Bilasipara (Assam) ...	2	0	0	1	12	7½	11·04
(3) Chandpur ...	2	0	0	1	13	1	9·18
Average ...	......			......			10·45

(6) Similarly some samples of wet jute were received from two *beparis* of Serajgunj. Like the samples of dry jute, these were also exposed to the sun for two days and then reweighed with the following results :—

Sample.	Original weight. lbs.	oz.	dr.	Weight after two days' drying in the sun. lbs.	oz.	dr.	Percentage of loss on drying.
(1) *a* Serajgunj ...	2	0	0	1	8	2½	24·61
b ,, ...	2	0	0	1	8	1½	
(2) *a* ,, ...	2	0	0	1	8	6	23·97
b ,, ...	2	0	0	1	8	4½	
(3) *a* ,, ...	2	0	0	1	8	4½	22·70
b ,, ...	2	0	0	1	9	3	
(4) Chandpur ...	2	0	0	1	9	0	21·88
Average ...	......			......			23·29

In both the above series the dry weight was taken about 3-30 P.M. As will be seen later on, all these samples reabsorbed some two per cent. of moisture by the next morning and one or two per cent. more by being stowed away for a few weeks from atmospheric humidity.

(7) The third sample in the wet series was examined at short intervals to see how quickly the moisture left the fibre. It was weighed out about 2 P.M. into two parcels of 2 lbs. each. Both the parcels were reweighed at 3-30 P.M. and again at 4-30 P.M. with the following results :—

		Weight at 2 P.M.	Weight at 3-30 P.M.	Weight at 4-30 P.M.
		lbs. oz. dr.	lbs. oz. dr.	lbs. oz. dr.
Sample 3 (Wet jute)	a ...	2 0 0	1 10 2½	1 9 6
	b ...	2 0 0	1 10 7	1 10 0

Taking the average of the two samples, it will be seen that the loss of weight at the end of 1½ and 2½ hours amounted respectively to 17·82 per cent. and 19·73 per cent. Seeing that the sample contained no less than 21·97 per cent. of moisture and that out of this as much as 10 per cent. was acceptable to the mills and balers, an hour's drying would probably have quite sufficed to make it also acceptable. In other words, jute parts with moisture so quickly that the plea of difficulty in drying to explain the presence of too much water can never stand.

(8) The three samples of dry jute mentioned in paragraph (5) were stored away in a dry but shady place for a few weeks, after which they were examined with the following results :—

Samples.	17th December, 10 A.M.	17th December, 2 P.M. (after 4 hours' drying.)	18th December, 8 A.M.	23rd December, 8 A.M.
	lbs. oz. dr.	lbs. oz. dr.	lbs. oz. dr.	lbs. oz. dr.
(1) Dry Mymensingh	1 14 1	1 13 2	1 14 1	1 14 1
(2) Do. Bilasipara ...	1 14 1	1 13 1	1 14 0	1 14 1½
(3) Do. Chandpur ...	1 14 5	1 13 5½	1 14 2	......

It will be seen that none of the samples returned to their original weight by reabsorption of moisture after they had once been thoroughly dried in the sun. The "bone-dry" fibre does gain a little (between 3 to 4 per cent.) in weight with the fall of temperature and a consequent rise in the humidity of the atmosphere, but not more, however long the fibre may stand. The permanent loss in weight by bone-drying in the case of the dry samples was about 6 per cent., and in the case of the badly watered samples as much as 18 to 20 per cent.

9. Nor is it any excuse to say that jute gets damp by exposure to night dew. Three experiments were made with bone-dry jute (1) by spreading out two samples to the dew at night, (2) by leaving by their side two other samples in bundles exposed to the dew, and (3) by leaving yet two other samples in bundles under a thatch. A fourth sample was kept inside a room. The following results were obtained in the morning:—

Nature of treatment.	Gain in weight by absorption of moisture.
	Per cent.
(1) Opened out to night dew ...	18·52
(2) Left in bundles exposed to dew ...	6·40
(3) Ditto under a thatch ...	2·93
(4) Ditto inside a room ...	2 42

The third experiment shows that a simple thatch would be a very good protection and keep the jute thoroughly dry. Jute therefore can not only be very easily dried, but can also be as simply kept dry. Excessive moisture almost in every case points to deliberate fraud.

10. A sample of dry Mymensingh jute that had already lost 5·85 per cent. was dried in a steam chamber. It was found that it lost 8·14 per cent. of its weight. So this sample contained almost exactly 14 per cent. of moisture. On the other hand, the absolute quantity of moisture in a bad sample of fraudulently-watered jute seems to be no less than 28 or 29 per cent. The absolute quantity of moisture in any sample of jute, however, is not a thing of much practical value, as the arrangements and apparatus, as also the skill required for its determination, are not at the command of practical men. But the relative amounts of moisture in a well-dried and a fraudulently-watered sample may be always ascertained very easily by almost anybody.

CHAPTER IV

Jute in Backergunge

(Written in 1898)

JUTE is cultivated throughout the district, excepting the tracts swamped by salt water, where it is grown on a very small scale in homestead lands for the home consumption only of the raiyats. According to the last jute forecast report, published in 1897 by the Agricultural Department, the normal area under jute in Backergunge is 6,400 acres only (37,623 acres in 1920, 50,000 acres in 1930).

Classification and Description

There are several races of jute, three or four races being commonly cultivated together under a common name. The names are confusing even in the same village. However, the following names adopted by the author signify particular races :—

(1) *Corchorus capsularis* :—
- (*a*) *Bagi*=*Bangi*=*Cowpla.*
- (*b*) *Raktabagi*=*Nileta.*
- (*c*) *Sutpat*=*Pechi.*

(2) *Corchorus olitorius* :—
- (*d*) *Bagi*=*Bangi*=*Nileta.*

The races may be distinguished by the colour of the stems and petioles of the plants, and the time of flowering and the normal height attained by the plants of a particular race. The colour of the plants (*a*) and (*d*) is light green. The plants of (*b*) are purple in colour, and the colour of (*c*) plants is reddish. We found there are two kinds of plants in each of (*a*) and (*d*). The petioles and the ribs of the leaves of the plants (*a*) and (*d*) are as green as the stems. But there are a few plants in (*a*) having their petioles and ribs of the leaves reddish,* and in (*d*) they are reddish or pinkish in the colour of the stalks, petioles and ribs.

---

* It has been observed that some of the races, having stems light-green and petioles reddish, change the colour of the stems to pinkish when old (as Amonia of Faridpur, Deswal of Serajgunj) ; but there are some (Dhupri of Faridpur, Bhadoya-white of Jalpaiguri. Deswal of Serajgunj) which are not so. In *C. olitorius* no such plants having reddish petioles on green stems are found. The same colour is contained by a plant.

The two principal species (*i.e., Corchorus capsularis* and *Corchorus olitorius*) differ from one another in the fruit. *Corchorus capsularis* has round pods, while the pods of *Corchorus olitorius* are long. They also differ in the ways of cultivation. *Corchorus capsularis* can thrive more or less in water to the depth of about 2 to 3 cubits, while *Corchorus olitorius* does not do so. Consequently *Corchorus capsularis* is very largely cultivated in Eastern Bengal, where the cultivation of *Corchorus olitorius* is but limited. It is also peculiar that *Corchorus olitorius* cannot strike the soil on which it is cultivated so firmly as *Corchorus capsularis* does.

The *Sutpat* is decidedly inferior to other varieties in the length of the fibre, but the cluster of fibre around the pith of *Sutpat* (*i.e.,* the bark) is thicker than that of the other three varieties. Therefore in yields of fibre all the varieties are nearly equal.

The fibres of the two species of *Bagi* are bright in lustre, hence they are called *Bagi* or *Bangi* (perhaps from *banga,* which is a provincial word for cotton, meaning white), which seems to have been derived from *bak* (stork), pronounced as *bag* in Eastern Bengal. Where both the species of *Bagi* are under cultivation *Corchorus capsularis* is called *Cowpla,* which might have been derived from *capor* (cloth), (*cowpla*=having the quality of *capor, i.e.,* as white as *capor*).

Rakta Bagi means red *Bagi.* The plants are purple, but produce white fibre. *Nileta,* from *nal* (red), is derived from the correct word *lal.*

I am unable to give reasons for the name of *Sutpat. Sut* means thread. Perhaps it means the *pat* (jute) which yields thread (*sut*). *Pechi,* meaning deformed, is attributed to plants which do not grow tall. There is no doubt that this variety was under cultivation in Eastern Bengal before all other varieties were introduced. I am also unable to give satisfactory reasons as to why *Bagi of Corchorus olitorius* is called in some places *Nileta.* With the exception of a few, the plants are not red (*nal*) in colour. It is probable that this name is derived from *nal,* a reed, or perhaps given to separate it from another species of the same name, *i.e., Bagi,* under *Corchorus capsularis,* which was first introduced. Where *Corchorus olitorius* was introduced before *Corchorus capsularis,* the latter is not called *Bagi,* but *Cowpla* as described above.

The opinion of the raiyats regarding the qualities of the varieties is not unanimous. The fibre of *Corchorus*

olitorius is said by the majority of them to be the stronger, but coarser. The colour of the fibre of this species is a little yellow, which is very much liked by the *Kapali* class of Eastern Bengal who weave gunny bags. The bags of *Corchorus olitorius* bring a higher price. The fibre of *Corchorus capsularis* is sometimes kept wet for two or three days after washing to give it this peculiar colour. This process would also give a sparkling yellowish colour to the fibre of *Corchorus olitorius.*

Hibiscus cannabinus (*Mestha* or *Mesot, i.e., Deccan* or *Ambari* hemp) is also included in jute in the market. It was largely cultivated on low lands a few years ago, when *mahajans* offered a higher price for it. Now the price for *Mestha* has much fallen off because the fibre is coarse. *Mestha* is generally longer than *Corchorus* and more easily stripped and washed.

The cause of Inferiority

Now I come to the main point of enquiry. The inferiority of the Backergunge jute is partly due to the mode of preparation of the fibre, but mainly to the conditions not being favourable for the crop.

Jute is chiefly steeped in four kinds of water; namely :—

(*a*) Stagnant *bheel* water.	(*c*) Running *bheel* water.
(*b*) Stagnant river water.	(*d*) Running river water.

The *bheel* water is called *kala jal* (black water) and the river water *baga jal* (white water). *Kala jal* is so called because it looks black (*kala*) on account of absence of silt. *Bheel* water is always clear and looks black. *Baga jal* seems to have been derived from *bag* (stork), which is white in colour. In fact, *baga jal* is light grey, on account of a large admixture of clay carried off by the rivers of the district which are the branches of the Ganges (Padma). It is so called because the raiyats do not generally make any difference between white and light grey colours.

The fibre of the jute steeped in the " black water " gives it a glossy white colour, while the colour of the jute steeped in *baga jal* is grey. The cause is this, as explained above, the *baga jal* contains silts which deposit in the fibre of jute at the time of retting. This silt colours the fibre grey. The stagnant *baga* water, of course, gives jute a better colour because it does not contain much silt. In the case of " black water," whether running or stagnant, the colour of the jute is not affected. The steeping of jute

in running water takes a longer time in retting than in stagnant water. The constant washing of the bundles by running water prevents an earlier decomposition. On the other hand, the bundles within (several bundles are heaped together) ret earlier than the outside bundles. Thus if the retting of the bundles within is complete, the bundles around will remain under-retted; and if the outside bundles are properly retted, the inside bundles will be overdone. Raiyats generally take the heap away when the bundles within are considered to be fit for stripping. The fibre of the bundles around will be specky and of bad quality. The fibre should be washed in running water instead of stagnant water, so that the refuse left after washing could be carried off immediately, leaving clean water for washing the remaining fibre.

Jute steeped in clean stagnant water, properly retted and washed in clean running water, is the best. Places having all these advantages are best suited to jute cultivation. The cause of the specky jute in Backergunge and in other districts is mainly due to insufficient retting. Insufficient retting does not clear the gummy matter of the coat of the fibre. Raiyats sometimes prefer insufficient retting where the prices for the superior and inferior qualities are not much different. Sufficient retting causes some loss of the fibre at the top of a plant and washing of the fibre clears every refuse, making the fibre lighter in weight. We have seen in some places in Bhola that the raiyats do not wash the fibre at all in order to make the fibre heavier.

A superior quality of jute is obtained in the *bheel* lands in the Gaurnadi Police-station, and a medium quality is produced in the *chars* where jute is steeped in ditch water which is stagnant, though the ditch water even is filthy and brownish in colour with decayed vegetation. The quality is inferior in all places by the sides of rivers where it is steeped in running *baga jal.* It does not appear to us that brackish water has anything to do with the inferiority of the Backergunge jute; in fact there is no brackish water during the rains in the jute-growing tracts of Backergunge.

CHAPTER V

Races of Jute in Pabna

(Written in 1899)

It was observed that there are some particular races which yield fibres of very good quality. Structure of soils and situation of lands have a great deal to do with the quality. The same variety yields fibres of different qualities in different places. Sandy loam is more adapted for the crop than either clay or sandy soil as far as the quality is concerned. Clay soil is said to produce the heaviest outturn. Flood also affects the quality of the fibre. The colour of the fibre depends chiefly on the cleanness of the water for steeping and washing.

The time of flowering has not been mentioned in the case of each race. Early races commonly run to flower in August, and the late races in September. The same race flowers at different times at different places, even at the same place if grown under different conditions.

The following races of jute were separated by the author from the mixed plants of several races as cultivated together by the raiyats. The specimens of selected plants of particular races Nos. 1 to 8 were collected by him at Serajgunj during 1899 for the Reporter on Economic Products to the Government of India. They are described as follows :—

Species—*Corchorus capsularis.*

No. 1.

It is locally called *Desal* (country).

Description.—Stem and leaf-stalk reddish (purplish or pinkish). It flowers in August.

Outturn.—Five maunds per bigha (3 bighas = 1 acre).

No. 2.

Called *Desal,* separated from the same field as noted under No. 1.

Description.—Stem light green; leaf-stalk and fruit reddish.

These races of *Desal,* as described under Nos. 1, 2 and 3, are always cultivated together.

No. 3.

It is also called *Desal.*

Description.—Stem and leaf-stalk are light green.

No. 4.

It is locally called *Kakya* or *Bombai* or *Kakya Bombai*. *Bombai* is perhaps so called as this variety is taller than others. The word *Bombai* (Bombay) is used in some places in Lower Bengal to denote the agricultural products which are larger in size. We could not ascertain what they meant by *Kakya*. Probably the name is derived from the word Kayia, vulgarly applied, in Bengal, to the Marwaris, who are well-known painstaking traders. It is possible that the Kayia traders introduced this jute into Serajgunj.

Description.—Stem, leaf-stalk and fruit are light green. It flowers about one month later than the *Desal*. It yields, it was said, more than *Desal* by about 50 per cent. It also resists water-logging better than the *Desal*. It can grow in water to the depth of 5 or 6 feet, whereas the *Desal* does not grow in water more than 3 or 4 feet in depth. Its bark is thicker. The fibre is more glossy than that of *Desal,* though the fibre of *Desal* is finer. Raiyats like this variety most, because its fibre is sold at higher prices. The seed of this variety is yearly indented from Assam. It is less branchy than any other race of *C. capsularis*. The raiyats of the village from where the specimens were collected did not like to waste fibre of such quality by allowing the plants to seed. It should also be mentioned that the Serajgunj area, consisting mostly of low lands, was not suitable for producing jute seed. In future the Agricultural Department should indent the seed of this variety for distribution, especially in flooded districts.

Two other races were found in the same field cultivated with the *Kakya Bombai* jute. One of them had a reddish stem, while the other had a light green stem with reddish petioles. It flowers later than the *Desal* races; otherwise the mixed *Kakya Bombai,* as cultivated, would have been the same as *Desal* of the locality. Taking all points into consideration, it resembles the green-stemmed plants of *Hewti* jute of North Bengal.

The difference between *Desal* and *Kakya Bombai* is in the following respects :—

(*a*) *Kakya Bombai* thrives better.

(*b*) The outturn of *Kakya Bombai* in Serajgunj is 9 maunds, and that of *Desal* 8 maunds. *Nileta* race (No. 5) yields 7 maunds per bigha, when the plants are all well grown.

(*c*) The *Desal* yields much more rooty fibres (adventitious roots) than the *Kakya Bombai.*

(*d*) The seed of the *Kakya Bombai* is a little smaller than that of the *Desal.*

No. 5.

Locally called *Nileta.*

Description.—Stem and leaf-stalk are a purple colour. It flowers about a fortnight later than the *Desal.* It grows well on dry land, but can thrive also in water up to three feet deep. The bark is thinner than that of *Desal.* The colour of the fibre is more glossy than the colours of *Desal* and *Kakya Bombai;* hence it fetches higher prices than the others.

Species—*Corchorus olitorius*

No. 6.

Locally called *Tosha.*

Description.—Stem, branch and leaf-stalk are of pink colour. It grows taller than the *Desal.* The fibre is coarser and more yellowish; hence it is sold at a lower price. It does not grow in water. It flowers about two months later than the *Desal.* Two seers of the seed of the *Desal* jute are sown per bigha, while *Tosha* seed is sown at the rate of one seer per bigha. The yield of fibre of the *Tosha* jute is somewhat heavier than that of the *Desal,* that is, if the *Desal* produces 8 maunds per bigha, the *Tosha* will produce 10 maunds. It may however be noted that *Tosha* is always cultivated on high land.

No. 7.

Locally called *Barpata,* which means long (*bara*), jute (*pata* or *pat*).

Description.—Stem and leaf-stalk are light green. It flowers about a fortnight earlier than the *Tosha.* The colour of the fibre is yellowish—less yellow than that of *Tosha.* In outturn Nos. 6 and 7 are similar.

No. 8.

It is called *Tosha.*

Description.—Stem and leaf-stalk are slightly reddish or pinkish. These specimens were collected from the field No. 6.

CHAPTER VI

Races of Jute in Mymensingh

(Written in 1899)

MYMENSINGH jute tops all other qualities in the market. Its superiority is chiefly due to the fact that the crop of Mymensingh is extensively cultivated on high lands and steeped and washed in clean water of the old Brahmaputra, and not to any particular race or races cultivated in the district. The following are the descriptions of the typical races separated and collected by us from fields at Jagannathganj.

Species—*Corchorus capsularis*

No. 1.

Locally called *Chhoran,* derived from the word *chhota* (small). It is a dwarf and early variety.

Description.—Stem, leaf-stalk and fruit of the typical plants collected are green. It flowers in *Ashar,* and the fruit ripens in *Bhadra* and *Aswin.*

No. 2.

Locally called *Baran,* which means big (*baro*).

Description.—Stem, leaf-stalk and fruit of the typical race are of light green colour. It grows taller than *Chhoran,* hence it is called *Baran.* Flowering and fruiting seasons begin later than the former. It produces fewer branches than the other. The outturn of the fibre is also heavier than that of *Chhoran.*

No. 3.

It is locally called *Kuchmadan* or *Nalpat* (red jute).

Description.—Stem, leaf-stalk and fruit are reddish. The fibres of the former two kinds are said to be better than the fibre of this variety, both in colour and quality. Flowering and fruiting time is the same as that of the *Baran* jute.

No. 4.

Called *Kamarjani.*

Kamarjani is a jute mart in Rangpur. The seed of this variety was perhaps first got from Kamarjani, hence it is so called.

Description.—Stem, leaf-stalk and fruits of the typical plants are light green. It is less branchy and somewhat taller than *Baran.*

It also produces a heavier outturn than the typical *Baran* race. It grows very well on dry land. This is one of the best varieties which we recommend for introduction into other districts. Flowering and fruiting seasons commence about one month later.

Species—*Corchorus olitorius*

No. 5.

Locally it is called by the names of *Nalita, Talla, Barpata, Baran.* All these words mean the same thing, which is tall. *Nalita* from *Nal* (reed). *Talla* derived its name perhaps from a kind of bamboo which is called *Talla bans.* Originally this word (*talla*) might have been derived from the word *Taral* (*Tarala*), a Sanskrit word which means liquid (light). This bamboo is much lighter than other bamboos.

Description.—Stem, leaf-stalk and fruit of the typical plants collected are light green. The fruit ripens in *Kartik* and *Aghran.* Its outturn is said to be heavier than that of the former races.

No. 6.

Local names.—*Nileta, Talla.*

Description.—Stem and leaf-stalk of the typical plants are pinkish. It is identical with the former in other respects.

CHAPTER VII

Races of Jute in Dacca

(Written in 1899)

DACCA is one of the most important jute districts in Bengal. The quality of the Dacca crop, which is produced in the area watered by the old Brahmaputra, is identical with the high grade Mymensingh jute. Specimens of the races of jute which were separated and collected by us from Narsingdi *Thana* (Police Station), for Dr. Watt, are described below :—

Species—*Corchorus capsularis*

No. 1.

Local names.—*Dhaleswari, Dhalsundar.* Both the words denote white plants.

Description.—Stem, leaf-stalk and fruit are light green. It flowers in July, and the fruit ripens in September and October. It is the variety most commonly cultivated in this subdivision (Naraingunj). It is a good variety.

No. 2.

Local name.—*Parbatya.* This name might have been derived from the word *parbat,* a hill. The seed was perhaps obtained from some hilly place in Tippera.

Description.—Stem, leaf-stalk and fruit pinkish; growth luxuriant; outturn heavier than that of the *Dhaleswari.* Fibre is also said to be better in quality and colour.

No. 3.

Local name.—*Agniswar,* from *agni*=fire. Here it means that it has the colour of fire.

Description.—Stem, branches and fruit are all along pink; leaf-stalk pinkish. Fibre is as good as *Parbatya.* Cultivation has been recently introduced.

No. 4.

Local names.—*Desi, Desal.*

Description.—The colour of the stem, branches and leaf-stalk is pinkish, like the colour of *Parbatya.* But in growth it is shorter and it ripens somewhat earlier. It is now rarely cultivated.

Species—*Corchorus olitorius*

No. 5.

Local name.—*Dewnallya.* Here *Nallya* denotes *Corchorus olitorius. Nallya,* perhaps derived from *nal,* a reed which is tall.

Description.—The colour of the stem, leaf-stalk and fruit is light green. It flowers in September. Other particulars are the same as given of *Corchorus olitorius* in other districts.

No. 6.

Local names.—*Bangi, Nallya.*

Description.—Stem pink or light red; branches and leaf-stalk are pinkish. It flowers in September. Fibre is more yellow than that of the former.

CHAPTER VIII

Races of Jute in Tippera

(Written in 1899)

TIPPERA comes next to Dacca as a district of high-class jute. *C. olitorius* is hardly cultivated in the district. Typical races, of which specimens were collected from Chandpur, are described below :—

Species—*Corchorus capsularis*

No. 1.

Local name.—*Bidyasundar.*

Description.—Stem and leaf-stalk are purple. Fruit purplish. It yields the finest and most glossy fibre of this place. Outturn 8 maunds per bigha (standard).

No. 2.

Local name.—*Dew Dholi. Dew* from *Dev*=a god, and *Dholi*=white. *Dew Dholi* means white like a god.

Description.—Stem, leaf-stalk and fruit are light green. It is second to the former as regards quality. But the outturn is somewhat heavier, *i.e.,* 10 maunds per standard bigha.

No. 3.

Local name.—*Dhalsundar,* meaning white, which is beautiful (*sundar*).

Description.—Main stem is light green; branches and fruit pinkish; it yields heaviest outturn, *i.e.,* 11 maunds per standard bigha. Fibre is said to be coarser than that of the *Dew Dholi.*

CHAPTER IX

Jute in Faridpur

(Written in 1900)

CORCHORUS OLITORIUS is always cultivated on high lands above inundation level. In some places we have found it growing in water only to the depth of a few inches, at about the time of harvest. It is sown in *Bysak* (middle of April to the middle of May). The yield per bigha is here about 7 maunds (sometimes 8 maunds). The raiyats are of the opinion that the fibre of *Corchorus olitorius* is stronger than the fibre of *Corchorus capsularis*. The bark of *Corchorus olitorius* is said to be thicker than that of *Corchorus capsularis*. The best fibre is obtained when the plants are just in flower in both the cases. The heaviest yield is obtainable when the fruits of the plants are mature. The *Bagipat* is not generally harvested before it bears fruit. Harvesting begins in *Sravan* (middle of July to middle of August), and continues up to the middle of *Aswin* (the end of September).

There are two methods in stripping jute fibre. They are as follows :—

(1) After the retting is complete, the plants are carried home. A plant is got hold of by the right hand of the stripper, who is generally a woman member of the family. She then pulls off the fibre by her left hand, and the stalk (wood) slides easily away. During this operation of the pulling off the fibre, she keeps her point finger between the fibre and the wood. One skilled woman can strip about 15 seers of dry fibre per diem.

(2) After the retting is complete, the raiyat stands by the heap of the plants. He takes up a handful of the plants which are broken at the distance of about 18 inches from the bottom. The broken pieces of stalks are then thrown away. He now takes hold of the separated fibre at the bottom by both the hands, and pushes the plants up and down. After five or six pushes, the fibre is cleared off. By this method one hard-working man can strip about 20 seers of fibre per diem.

In Faridpur both the methods are in vogue. In Madaripur and Sadar subdivisions. jute is stripped in the method (1) and in Goalundo in the method (2).

Deep, clear and stagnant water is preferred for steeping jute. Steeping in running water takes a longer time for retting.

Commercial Specimens of Madaripur Jute

At the Madaripur market there are four kinds of jute, *viz.*—

(1) *Charua*, (3) *Bilan*, and
(2) *Bagi*, (4) *Puban*.

(1) *Charua* belongs to the species *Corchorus capsularis.* It is grown on *char* lands on the sides of the rivers Padma and Arealkhan. Water to the depth of about two cubits accumulates on such lands, during the months of *Sravan* and *Bhadra.*

The soil of these places may be described as a loam inclined to be sandy. These parts of the district produce the best jute grown in the subdivision. It is steeped in deep stagnant river water. The fibre is available in the market from *Sravan* to *Bhadra.* It was sold at the time of our visit at Rs. 4-8 per maund. This good quality of the fibre is said to be due to the soil where it is grown.

(2) *Bagipat* (*Corchorus olitorius*).—It is grown on high lands and steeped in stagnant *bheel* water, and its fibre is available in the market from *Bhadra* to *Aswin.* It is well known to the merchants that it is coarser than *Corchorus capsularis,* but its rope is said to be stronger than that made of the latter. The price was at Rs. 4-7 per maund. The soil on which it is grown is generally sandy loam. Out of the total quantity of jute available at the market of Madaripur, *Bagipat* contributes only about 12 per cent.

(3) *Bilan* (from *bheel*) belongs to the species *Corchorus capsularis,* grown in *bheel* lands where water accumulates to the depth of about three cubits. All the varieties that are cultivated on low lands do not thrive much after inundation. It is steeped in stagnant *bheel* water. This quality of jute is available in the market from *Ashar* to *Aswin.* Price Rs. 4-4 per maund. The soil on which it is grown is clayey.

(4) *Puban* (*eastern*) is grown on *char* lands. The soil is similar to that on which *charua* is grown. It is steeped in shallow running river water (depth only $2\frac{1}{2}$ cubits). The mud of the river water deposits on the fibre during the time of retting, which spoils the colour of

the fibre. It was sold at Rs. 3-12 only per maund. It is available in the market from *Sravan* to *Bhadra*.

Commercial Specimens of Pangsa Jute

The following kinds of jute are sold at the Pangsa market :—

(1) *Pangsa*, (3) *Padma paira*,
(2) *Charua*, (4) *Dakhina*, and
(5) *Faridpuri*.

(1) The *Pangsa* quality of jute is brought for sale to Pangsa from surrounding places. It is sold in the market from the middle of *Sravan* to *Kartik*. This sort of jute is cultivated on high lands composed of loamy soil. It is steeped in clear water. The price was Rs. 3-5 per maund (1 seer=60 tolas). There is not less than 10 per cent. moisture, generally 30 per cent., in the fibres which are sold at this market. It belongs mostly to the species *Corchorus capsularis*.

(2) *Charua* is grown on *char* lands composed of a loam inclined to sands. It is steeped in stagnant river water slightly muddy. Fibre is available in the market from the middle of *Sravan* to *Kartik*. Price was Rs. 3-6 per maund. *Pangsa* and *Charua* are generally mixed together and sent to Calcutta for sale.

(3) *Padma paira* is a sort of jute which comes from the other side of the river Padma (*e.g.*, Pabna). It is grown on *bheel* lands composed of clayey soil. It is more or less specky and reddish. It is available in the market from *Bhadra* to *Kartik*. Price Rs. 3-4 per maund.

(4) *Dakhina*.—This sort of jute comes from the Balliakandi police-station. It is grown on *bheel* lands composed of clayey soil. It is steeped in stagnant clear water. Fibre is specky. It is available in the market from *Bhadra* to *Kartik*. Price Rs. 3-4 per maund.

(5) *Faridpuri* sort of jute is brought from places in the police-stations of Ainpur and Bhanga. It is grown on *bheel* lands composed of heavy clay. The water in which it is steeped is more or less clear. Fibre is very specky, and available in the market from *Sravan* to *Aswin*. Price Rs. 3 per maund.

Localities where jute of good quality is obtained sometimes produce a bad quality, and for the following reasons :—

Jute is sown from *Chaitra* to *Jaistha*. During the period of growth occasional showers of rain at intervals of about eight days are required for producing good

quality, but heavy rains during this period (when the plants are young) are injurious. Hairy larvæ, locally called *salkit,* also attack the plants if there be heavy rains; the plants become more branchy than usual, and yield a specky and harsh fibre.

The best fibre is obtained from plants which are harvested when just in flower. Early flood is therefore wanted by raiyats for harvesting (*e.g.*, steeping) just in time.

Races in Madaripur

Samples from Nos. 1 to 8 were separated and collected in Madaripur, on the 11th and the 12th September, 1900.

Species—*Corchorus capsularis*

No. 1.

Local name.—*Sutpat.*

Description.—Colour of the stem and leaf-stalk is pinkish. It flowers in the latter part of July. This variety is commonly cultivated in this part of the district, because it can thrive well in deep water (not exceeding 3 cubits in depth). Fibre is generally specky at the bottom. About 5 maunds of fibre are obtained from one bigha of land. It grows in deep water.

No. 2.

Local names.—*Desal* (country), *Peti* (small).

Description.—Stem is light green when young. Branches, leaf-stalk and fruit are of reddish colour similar to No. 1. It is being cultivated here for a long time. It is more branchy than other varieties. It grows in water to the depth of 5 feet, yielding 6 or 7 maunds per local bigha (60 × 60 yards).

No. 3.

Local name.—*Udhap* (meaning perhaps white = *dhop*).

Description.—Stem, leaf-stalk and fruit light green. Fruit larger than those of the *Desal.* Outturn 7 maunds per bigha (local).

No. 4.

Local name.—*Cowpla pat.*

Description.—All parts of the plant are light green in colour. It flowers in the latter part of July. It is grown on both high and low lands. Outturn is about 5 maunds per bigha, but the quality of the fibre is said to be better than that of *Sutpat.*

No. 5.

Local name.—*Lali* or *Ranga pat.*

Description.—The colour of the stem and the leaf-stalk is purple. Grown in both high and low lands. It flowers about a fortnight earlier than the races 3 and 4 mentioned above. Fibre is said to be finer but less in quantity.

No. 6.

Local name.—*Belgachi.*

This race has been newly introduced.

Description.—Stem, leaf-stalk and fruit are pinkish or reddish. It flowers in the latter part of July. Fibre is better than that of the *Sutpat,* both in quality and quantity.

Retting is finished in much shorter time. In June it takes only 10 days for retting, while *Desal* or *Sutpat* takes 15 days. In September it takes 15 days.

Species—*Corchorus olitorius*

No. 7.

Local name.—*Bagi.*

Description.—All parts of the plant are light green in colour. Generally sown in May. Outturn 8 maunds per local bigha (one bigha = 120 × 120 cubits = $\frac{3}{4}$ acre).

No. 8.

Local names.—*Lal, Nal* or *Ranga bagi.*

Description.—All parts of the plant are pink in colour. No distinction is made between these two races of *Corchorus olitorius.* Regarding the cultivation, yield and quality, they are identical.

Races in Goalundo

The following 9 races, *i.e.,* from No. 9 to No. 17, were collected from Pangsa, the principal jute market in the Goalundo subdivision in Faridpur, and those from No. 18 to No. 23 from Goalundo.

Species—*Corchorus olitorius*

No. 9.

Local names.—*Satnala, Barpata, Dhamraj.*

Description.—Stem, leaves and pods are all light green in colour. Sown in *Chaitra* and *Bysak* on high lands. Seed sown, 1 seer (60 tolas) per *pakhi.* Flowers in *Bhadra.* Outturn per *pakhi* is 6 maunds to 7 maunds (1 *pakhi* = 10 × 12 *nals,* and 1 *nal* = $6\frac{1}{2}$ cubits = ·8 standard Bengal bigha). Harvested in *Bhadra.* It takes 12 days for retting.

No. 10.

Local name.—*Meghnal.* It is also called *Barpata* in some places.

Description.—Stem and leaf-stalk are of pink colour. Sown in *Chaitra* and *Bysak* at the rate of 1 seer (60 tolas) per *pakhi.* Flowers in the latter part of *Sravan.* Outturn per *pakhi* is about 6 maunds. The plants are a little shorter than the former. Harvested in *Bhadra.* It takes 12 days for retting. *Corchorus olitorius* grows always taller than *Corchorus capsularis.*

Species—*Corchorus capsularis*

No. 11.

Local name.—*Dhaleswari.*

Description.—Stem, leaf-stalk and fruit are reddish in colour when old. It is sown in *Chaitra* and *Bysak* at the rate of 2 seers of seed per *pakhi.* Flowers early in *Sravan.* The bark is thinner than that of *Corchorus olitorius.* It may be grown on both high and low lands. Water accumulated on the low lands should not be more than $1\frac{1}{2}$ cubits in depth. It takes about 15 days for retting.

Outturn.—Four to $4\frac{1}{2}$ maunds per *pakhi.*

No. 12.

Local name.—*Amunia.*

Description.—Stem, leaf-stalk and pod are light green in colour, and it does not branch much. Sown on high lands in *Chaitra* and *Bysak.* Two seers of seed are required for one *pakhi.* Flowers in the latter part of *Bhadra.*

Outturn.—Six maunds per *pakhi.*

It takes about 12 days for retting. The fibre of this variety is the finest of all grown in this part of the district.

No. 13.

Local name.—*Shani;* also called *Desal, Sodu* and *Kuch mardan.*

Description.—Stem, leaf-stalk and pod are reddish. Sown in *Chaitra* and *Bysak* on both high and low lands. Water on the lands here stands to the depth of about 3 to $3\frac{1}{2}$ cubits. Seed rate 2 seers (1 seer=60 tolas) per *pakhi.* Flowers early in *Ashar.*

Outturn.—Three to $3\frac{1}{2}$ maunds of fibre per *pakhi.*

It takes about one month for retting.

This race has been cultivated here for a long time. Fibre is always specky at the bottom. Plants are very strong and much branchy. Stripping is very difficult. Its cultivation is dying out.

No. 14.

Local name.—*Dhapa.*

Description.—Stem, leaf-stalk and pod are reddish, identical with No. 13, but it flowers in *Bhadra.*

No. 15.

Local name.—*Meghnal* (see No. 10).

Description.—Stem and leaf-stalk are of purple colour. Sown generally on low lands where water accumulates to the depth of about 2 cubits. One and three-quarters of a seer (60 tolas = 1 seer) of seed required for one *pakhi.* It flowers in early *Bhadra.* It takes about 15 days for retting.

Outturn.—About 5 maunds per *pakhi.*

No. 16.

Local name.—*Kamarjani* (1). It is also called *Parbatya.*

Description.—Stem, leaf-stalk and pod are light green in colour. Sown generally on low lands in *Chaitra* and *Bysak.* Water may accumulate there to the depth of about 2½ cubits. Two seers of seed (60 tolas = 1 seer) are sown per *pakhi.* Flowers in the latter part of *Sravan.* It takes about 20 days for retting.

Outturn.—Five maunds per *pakhi.*

Leaves of this race are used here as a favourite pot-herb.

No. 17.

Local name.—*Kamarjani* (2).

Description.—Stem and leaf-stalk are reddish. Sown together with No. 16.

Seed.—Two seers per *pakhi.* Flowers in the latter part of *Sravan.*

No. 18.

Local name.—*Ghagri.*

Description.—The colour of the stem, leaf-stalk and fruit is light green. It flowers in July, and the fruit ripens in September.

No. 19.

Local names.—*Ghagri, Desal.*

Description.—The colour of the stem is light green, and that of the branches, leaf-stalk and fruit is reddish. The fruit is a little smaller than that of the *Ghagri,* No. 18.

No. 20.

Local names.—*Lalpat* (red jute), *Kajla* (black), *Meghnal* (*Megh*=cloud, and *Nal*=red, *i.e.*, a colour mixed of blue and red).

Description.—Stem, branches and leaf-stalk are pink; a tinge of violet colour on the fruit. Flowering season commences about 15 days later than that of *Ghagri*. It is less branchy than the *Ghagri*. As a pot-herb, the leaves of this race are said to be better than the leaves of the other races.

No. 21.

Local name.—*Kamarjani*.

Description.—Stem, leaf-stalk and fruit are similar to No. 17. The fibre is more glossy than that of the *Ghagri*. It is one of the best varieties, recommended for introduction into other districts where it is not cultivated.

No. 22.

Local name.—*Dhupri* (perhaps from Dhubri in Assam). A new race just introduced.

Description.—Stem light green. Leaf-stalk and fruit are slightly reddish. Growth is more vigorous than that of other races of *Corchorus capsularis*. Flowering time begins some 15 days later than that of *Ghagri*.

Species—*Corchorus olitorius*

No. 23.

Local name.—*Bomi* (from *Bangi* or *Bagi*).

Description.—The colour of the stem and leaf-stalk is light green. It flowers in July. It does not grow as tall as *Satnala,* which is also a light green variety flowering later.

CHAPTER X

Races of Jute in Rangpur

(Written in 1900)

RANGPUR is the most important jute district in North Bengal. *C. capsularis* is commonly cultivated, hence it is called *Jati-pat,* meaning true jute. *C. olitorius* is scarcely met with in the district. North Bengal jute is fine in texture, as it is the product of high lands, but the general practice of over-steeping harms the quality to a certain extent.

Numerous names of jute by which the same races are denoted in different parts in East Bengal—nay in the same villages—are conspicuously diminished in the northern districts. The races are divided into two classes, *viz., Ausa* or *Bhadoya* (early) and *Hewta* (late). Each of the classes comprises various races. The *Ausa* and *Hewta* are, of course, cultivated separately. The *Ausa* races are preferred.

The following are the typical races of jute grown in Rangpur :—

Species—*Corchorus olitorius*

No. 1.

Local name.—*Parbati Madai.*

Description.—Stem and leaf-stalk are reddish in colour. There is another race under the same name of *Parbati pat,* the stem and leaf-stalk of which are light green. They are grown together.

Corchorus olitorius is sown during the months from *Falgun* to *Bysak.* In some places it is sown in *Bysak.*

It grows on high lands containing sandy loam. One seer (1 seer=90 tolas) of seed is required for 1 *done* (1 *done*=13 standard cottahs or 0·65 *bigha*).

Outturn.—Four maunds to 5 maunds fibre per *done.*

Flowering season.—Latter part of *Aswin.*

It is harvested when the plants are in flower. In the case of urgency, it is harvested just before the flowering time, but the yield per *done* would be only 3 maunds, instead of 4 maunds.

Ten to 15 days are required for retting.

It is said to be stronger but coarser than *Corchorus capsularis* jute.

It is only cultivated for home consumption, which may be roughly estimated at 3·5 per cent.

The price of the fibre of *Corchorus olitorius* is about 50 per cent. less than that of *Corchorus capsularis.*

Species—*Corchorus capsularis*

No. 2.

Corchorus capsularis is commonly called here *Jati-pat* (true jute).

Local name.—*Ausa. Ausa* is a crop which is harvested earlier than the *Hewta* jute. *Hewta* is from the word *Hementa,* the autumn season.

Description.—Stem is light green, but branches and leaf-stalk are reddish. Sown from *Falgun* to *Bysak* Seed rate one seer per *done.*

Soil.—Both high and low lands; but clayey soil is preferred, which is nothing but a loam.

Flowering season.—Latter part of *Sravan.*

It is harvested in flower and also in fruit.

Outturn.—Four maunds per *done.*

No. 3.

Local name.—*Ausa* (red). Also called in some places *Meghnal.*

Description.—Stem, leaf-stalk and fruit are purple in colour. In other respects than its yield of fibre, which is said to be a little less, *e.g.*, about $3\frac{1}{2}$ maunds per *done* instead of 4 maunds, it is similar to No. 2.

No. 4.

Local name.—*Hewta* (red), in some places *Meghnal.*

Description.—The colour of stem, etc., is reddish.

Sowing time.—*Falgun* to *Bysak.*

Flowering time.—Latter part of *Bhadra.*

Regarding the outturn, the opinion of some raiyats is that the *Hewta* variety yields heavier than the *Ausa* variety (about half a maund heavier per *done*); but others are of the opinion that the yields of the *Ausa* and the *Hewta* varieties are nearly equal. Some raiyats stated that the fibre of the *Hewta* races is more glossy than that of *Ausa.*

Regarding soil, cultivation, etc., it is similar to those of *Ausa* races.

No. 5.

Local name.—*Hewta* (green).

Description.—Stem and leaf-stalk are light green in colour. In other respects, it is similar to the variety No. 4, *Hewta* (red). But it is stated that a green-stemmed variety yields a little heavier outturn than any of the red-stemmed races. The red-stemmed races are however said to produce finer fibre.

The Rangpur raiyats prefer running water for steeping jute, as this water carries away all dirty substances from the heap of plants, although they are aware that it takes a longer time for retting. Best colour of the fibre is said to be produced if retted in such water. Jute is stripped in this district in the method No. 2 described in Chapter IX, under this Part of the Book.

CHAPTER XI

Races of Jute in Jalpaiguri

(Written in 1900)

JUTE is grown on high lands containing sandy loam.

Two seers of seed are sown per *done* (1 *done* is nearly equal to 1 standard bigha). It is sown from the 15th of *Chaitra* to the 15th of *Bysak* (*i.e.,* April) and harvested when the plants are in flower or in fruit. The raiyats are of the opinion that the harvesting operation of the plants in flower is easier and less expensive, while the plants harvested in fruit produce a little heavier outturn of fibre. In stagnant water retting is complete within a fortnight, while in running water it takes about a month; yet, running water is preferred because it gives a good colour to the fibre.

Yield is said to be only 3 maunds per *done.*

Species—*Corchorus capsularis*

Nos. 1 to 4, collected in Alipur Duars, and Nos. 5 to 9, at Jalpaiguri.

No. 1.

Local name.—*Betre.*

Description.—All parts of the plant are light green in colour. It flowers early in *Sravan.*

No. 2.

Local name.—*Betre.*

Description.—Stem is of light green colour, but branches and leaf-stalks are slightly reddish.

No. 3.

Local name.—*Betre.* It is also called *Parbatya.*

Description.—Stem and leaf-stalk are of purple colour. These three races under the name of *Betre* are always cultivated together in a mixture.

No. 4.

Local name.—*Hewti.*

Description.—All parts are light green. It flowers in *Aswin.* Fibre is said to be finer than that of the *Betre.* Extracting of fibre of the *Hewti* races is considered to be easier because *Hewti* is less branchy than the *Betre.* In other respects it is similar to *Betre.* Under the name of *Hewti,* there are also red-stemmed races, cultivated as usual in a mixture.

Corchorus olitorius is not cultivated in the Alipur Subdivision.

No. 5.

Local name.—*Bhadoya* (early). *Bhadoya* and *Betre* both denote early races.

Description.—Stem is light green, but leaf-stalk and fruit are slightly reddish. Sown in the months of *Chaitra* and *Bysak* on both high and low lands which are composed of sandy loam. It flowers early in *Sravan* and is harvested when the plants are in flower or in fruit.

Outturn.—Four maunds per *done.*

No. 6.

Local name.—*Bhadoya* (early).

Description.—All parts are of purple colour. *Lal pata* (red jute) is said to be better in colour and quality than the *sada pata* (white jute). In other respects it is similar to variety No. 5.

No. 7.

Local name.—*Hewti* (late).

Description.—All parts of the plant are light green. It flowers early in *Bhadra.* It is a late variety. The early varieties (Nos. 4 and 5) are said to yield better and heavier fibre. The bark of the plant of this variety is said to be thinner than that of the early varieties. In other respects they are similar. Early varieties take 15 days for retting, while late races take 20 days.

No. 8.

Local name.—*Hewti* (late).

Description.—Stem, leaf-stalk and fruit are of purple colour.

No. 9.

Local name.—*Hewti* (late).

Description.—Stem, leaf-stalk, etc., are of slightly purple colour or reddish.

Species—*Corchorus olitorius*

No. 10, collected at Jalpaiguri.

No. 10.

Local names.—*Marua pata, Bomochak.*

Description.—Stem and leaf-stalk are of pink colour. It flowers early in *Bhadra*. It is cultivated on a very small scale, usually for using the leaves as a pot-herb. It is stated that dried leaves of *Corchorus capsularis* are more efficacious for medicinal purposes. The leaves of *Corchorus olitorius* are not so bitter as those of *C. capsularis*, hence it is more palatable.

The fibre of this species is never available in the markets.

In Jalpaiguri, jute is stripped in the method No. 2 already described.

CHAPTER XII

Jute in Purnea

(*Published in* 1917 *in the Agricultural Journal of the Department of Agriculture, Bihar and Orissa*)

Area

PURNEA is the most important jute-growing district in the Province of Bihar and Orissa. According to the Jute Forecast Report of 1916, the area under jute in Purnea was estimated at 197,027 acres (140,000 acres in 1930). In 1915 the area under jute was reported to be 158,830 acres, while it was 281,220 acres in 1914. The heavy decrease in acreage in 1915 was due to the low prices obtained in 1914 after the outbreak of the great European War. On account of the rise in the price in 1915 there was an increase in 1916. On an average the normal area under jute in Purnea was then estimated at about 240,000 acres, that is, about 14 per cent. of the total cultivated area. The area under jute in 1916 included about 6,000 or 7,000 acres under *Hibiscus Cannabinus*, called *Amla patua*, cultivated in the Sadar Subdivision.

Soil

The soil of the district, excepting a small tract of old alluvium about a mile long and a half mile wide near the Monihari Railway Station, is new alluvium, mostly of light character. Heavy clay soil is met only in the *bheel* tracts in the Sadar Subdivision. The new alluvial soils are most suitable for the cultivation of jute. The *bheel* tracts are overflooded each year just after the break of the monsoon. Jute is sown early in these tracts and harvested in the beginning of the rainy season, before the crop is properly matured for cutting. By the middle of July about three-fourths of the crop in such tracts situated chiefly in Barsoi and Kadwa *thanas* are harvested. The high land jute in other parts of the district is of a superior quality. Jute is very much affected by water-logging when it is young. Although the plants of *C. capsularis* can bear water-logging admirably after they are in flower, water-logging is always harmful to the quality of the fibre.

Rainfall

The rainfall of the district during the growing period, that is, from March to August, averages 50 inches. This is quite sufficient for the crop, but sometimes heavy abnormal falls are received in July and August which cause great damage. For sowing, from 1 to 2 inches of rain are necessary for the preparation of the soil. After a fortnight, 1 to 2 inches of rain each week would be very favourable. Rain is necessary within a fortnight after sowing, otherwise the plants will begin to wither up. Heavy falls of rain, 4 to 5 inches at a time, within two months from the date of sowing will also damage the plants, and also interfere with weeding.

Rotation

As a rule, jute is sown on the same land, year after year, by the cultivators of small holdings. Well-off cultivators would sow paddy and jute in alternate years on the same plot. They know that jute will grow very well on the land after a lapse of 2 or 3 years. Jute is followed commonly by mustard in the *Rabi* season in the same year. Sometimes wheat, barley, pulses, also *San* hemp follow jute. Transplantation of paddy in the same land just after cutting jute is a common practice in the district. This is certainly an exhausting system of cultivation, especially where no artificial manure is applied. The value of sowing *San* hemp previous to the cultivation of jute is well known in the district. The *Rabi* crops, which are harvested late in February or in March, should never be sown on the lands intended for the jute crop, as it interferes with the preparation of the soil. An ordinary raiyat who has only a limited area will, however, cultivate jute on the same land, year after year, knowing fully that it will give him a poor outturn.

Manuring

Jute land is hardly manured in the district. The farmyard manure of a cultivator is reserved for tobacco only. It is known to the cultivator that cowdung is very good for jute, but he cannot afford to apply it to the crop after his tobacco field has been manured with it. It has been found that the field manured with about 8 or 9 cartloads of cowdung (about 6,000 pounds) would increase the yield by about 30 per cent. The next best manure for jute is castor-oil cake, which should be applied at the rate of about 480 lbs. per acre. Castor-oil cake is as yet unknown

in the district. The cost of 6,000 pounds (75 maunds) of cowdung would not be more than Rs. 4 or 5 if available for sale, while the castor cake at the rate of 480 lbs. to the acre would cost at least Rs. 12. Cowdung is therefore cheaper when obtainable.

The cultivation of the land for sowing jute begins in February, following a shower of rain, after cutting mustard or paddy. About six ploughings and cross-ploughings followed by *henga** are necessary to have a fine tilth of the soil. The seed will not germinate properly on a rough soil. In the low-lying lands sowing is finished in March, but on the high lands it extends up to the end of May. Early sowing is safer. The jute sown late in May, in the year 1916, was greatly damaged, if not altogether lost, in most places, on account of the early break of the monsoon, on the 8th June. Sowing should be finished by the middle of May.

Seed rate per acre in the district varies between 10 and 12 lbs.; but 9 pounds of seed would be quite sufficient. The seed is sown broadcast over the cultivated field after harrowing. Another harrowing covers the seed. If there is sufficient moisture in the field, the seed germinates in 3 or 4 days. After the plants grow 5 or 6 inches high, 2 *bidas* (rakings) at intervals of a week are given for stirring the soil as well as for thinning out the crowded seedlings. Afterwards the field is weeded once or twice before the rains set in. At the time of weeding the plants are again so thinned as to keep them a space of about 6 inches apart. This is done by the more intelligent cultivators. Some raiyats do not pay sufficient attention to thinning. In such cases the plants become thin and weak, yielding a poor outturn. In some parts of the district the raiyats will not take the trouble of removing the weeds from the field, but heap them at intervals thereon. This practice causes considerable gaps in the fields and thereby the raiyats lose some fibre unnecessarily. These heaps of uprooted weeds also become a source of future trouble from insect pests.

The plants grown on well-prepared and manured fields and treated properly afterwards will often grow to as much as 12 feet in height. On the average the plants of the district are 7 to 9 feet long. The longer the plants the higher will be the outturn of the fibre. Good soil and proper cultivation are, of course, essential for the success of the jute crop.

---

* A *henga* is a wooden beam used for crushing clods and levelling the field.

Varieties or Races

Corchorus capsularis is principally cultivated in the district of Purnea. It is known by the names of *Bhadoya* and *Amonia Patua,* of which *Bhadoya* (early jute) is most common, while *Amonia* (late jute) is cultivated to a very limited extent. There are principally three colours in the stems of the jute plants in both the early and late races, accordingly they are grouped as :—

(*i*) Green-stemmed plant called *safed* or *ujra patua.*
(*ii*) Pinkish-stemmed plant named *Meghnal.*
(*iii*) Purple-stemmed plant named also *Meghnal.*

Plants of all these three colours are found in mixture in one field. Regarding the yields, some raiyats say that *safed patua* is superior to the *Meghnal,* while others hold a contrary opinion. However, the plants of the *safed* races cultivated in the district branch more than the plants of the *Meghnal* races do, so it is easier to strip the fibre from the *Meghnal* jute. Anyhow, about 75 per cent. of the jute plants of Purnea belongs to the *safed* race.

Occasionally *C. olitorius,* known as *Mooniashi patua,* is cultivated in the homestead lands as a pot-herb and not usually as a fibre crop. The lands of the district are more or less subject to inundation and are not suitable for this jute. It cannot bear water-logging to any extent. Hence it is not possible to introduce this jute into the district. The *Kakya Bombai* jute of Bengal of the *Capsularis* species has been cultivated experimentally in some places. It has done well—better than the local varieties—but it is a late-maturing race, being harvested about a month later, and as most of the raiyats want to transplant paddy on the same land after cutting jute in July, so they must have an early race to sow. High lands, which are unsuitable for transplanted paddy, may be found suitable for this late jute. Undoubtedly *Kakya Bombai* is superior to the local races of jute in yield. Those who wish to cultivate the crop on a large scale should cultivate both early and late varieties, so that the time of harvesting might be extended considerably. It would not be possible for them to manage the harvesting of all the fields at the same time if only the early or late races be cultivated. Such cultivators should introduce the *Kakya Bombai* jute.

The fibre of *Hibiscus cannabinus* is sold as jute in the district. This fibre is a little coarser than jute, but it is not easy to separate it if mixed together. There is a considerable area under *Hibiscus* in this district.

Cutting

Cutting of jute commences from the beginning of July and continues till the end of September. As soon as the flowers appear, the plants are considered to be fit for cutting. The raiyats have to wait, however, till the neighbouring tanks or pools are filled with the rain or flood water, in order to steep the crop within an easy reach. If steeping water is not available in the neighbourhood of the fields, the cutting is usually delayed. The jute cut in flower is considered to be the best in quality, but it is known to give a higher yield if it is cut later on. Jute is hardly ever cut at the stage when the fruit of the plant is ripe, as it is difficult then to separate the fibre. In Kissenganj and elsewhere, where paddy is transplanted in the same field in the same season, the cutting is finished by the middle of August, at the latest.

The plants are cut to the ground with a sickle and then stacked in rows, one layer over another. The object of this is to remove as many leaves as possible. If the plants are steeped immediately after cutting, the leaves are said to discolour the fibre. In low lands, where the fields are overflooded, this shedding is not done. Usually after two or three days the leaves shed and the plants are made into bundles and carried to the nearest pools or ditches for steeping and retting. The ditch or pool water is commonly muddy, which discolours the fibre to some extent.

Steeping or Retting

The bundles are stacked in water, one layer over another, two or three layers in a stack which, in size, is about 10 or 12 feet square. This stack is commonly covered over with water weeds. The plants take 10 to 20 days to ret. If the plants are cut in the latter period of the season, it may take even a month for retting. In a hot temperature the retting is quick. The plants, cut in flower, take a shorter time than the plants in fruit. After a week or so, the plants of the stack should be frequently examined, in order to ascertain whether they are fit for stripping, as over-retting spoils the quality of the fibre.

Stripping and Washing

Separating the fibre from the stem, after the retting process is complete, must be finished within a couple of days. The stripper stands in water about 2 feet deep, takes a handful (*muthha*) of the stems and beats the

bottom of the same with a wooden flat *hata*, resembling a dairy hand. After a few beatings with this *hata* the fibre of the stems at the bottom becomes loose. The stems are then broken off about 1 foot from the bottom. The separated fibre is taken hold of by both hands and the stems are pushed forward and backward on the surface of the water. After a few jerks the fibre is cleared off the stalks. A skilled man can strip about 20 seers (40 lbs.) of fibre in a day, working 6 hours. An untrained man, however, cannot strip more than 10 seers (20 lbs.) of fibre. After separating the fibre from the stalk, the stripper washes the fibre by pulling it right and left in the same water. The colour of the fibre is greatly improved if it were washed in clean water afterwards. The cultivators, however, will seldom take the trouble of carrying the fibre to any other place for washing. The colour of the jute of this district is, in consequence, blackish, owing to steeping in muddy ditch water. After washing, the fibre is carried home and dried in the sun over bamboo poles. It is dry in two or three days, and then bundled for the market.

Outturn

The average outturn of jute of the district is estimated at 15 maunds (1,230 lbs.) per acre, if weather conditions are favourable

Trade

There are two principal classes of jute in the district, *viz.*, long and short. Jute which is over 8 feet in length is considered long jute, while that shorter than 8 feet is called short jute. About two-thirds of the jute of the district is long jute and the remaining portion is short. It is of a blackish colour. The jute of the inundated low-lying tracts of the district is barky (specky) at the bottom. The jute imported from Nepal, called *Morang* jute, is the best fibre on the market, as it is long and barkless. Between 3 and 4 lakhs of maunds of jute are imported into the district from Nepal annually.

Jute is commonly sold by the raiyats at their own places to the middlemen known as *beparies*, who take the fibre to the markets to sell to the *goladars*. Almost every *goladar* has a hand power press, with which he presses the jute into bales. These are called the *katcha* bales of the trade weighing 100 or 280 lbs. each. Reliable *goladars* press the long and short jute separately. The bales are sent to Calcutta. The chief jute markets in the district are

Kissenganj, Forbesganj, Kasba and Barsoi. At Kissenganj about 8 lakhs of maunds,* at Forbesganj about 6 lakhs of maunds, at Kasba or Barsoi about 3 lakhs of maunds of jute are sold, if the yield of the district is normal.

Besides the above-mentioned important markets, every railway station is also a jute market of more or less importance.

The local weights are different at different places. From 60 to 90 tolas make a seer. In most places 90 tolas are taken for a seer. One English pound is equivalent to 41 tolas, or roughly 40 tolas make a pound. Forty seers, either of 60 tolas or 90 tolas, make a maund.

Preservation of Seed

There is no one in the district who cultivates jute to obtain the seed for sale. The cultivator usually keeps for seed a portion of a plot of jute of stunted growth which is estimated to be sufficient for his own cultivation. An acre of land produces about 9 maunds to 10 maunds of seed.

Costs

The costs for the cultivation of an acre of land under jute in Purnea were calculated at about Rs. 65. The cultivators never keep any account to show their actual expenses for the cultivation. However, a table showing estimated costs for one acre of land is given below :—

	Rs.	As.	P.
8 ploughings (1 ploughing requires 3 ploughs) at 2 ploughs for a rupee	12	0	0
5 seers (10 lbs.) of seed	1	4	0
2 *bidas* for raking and thinning ...	1	0	0
1st weeding, 24 men, at annas 6 per head	9	0	0
2nd weeding, 12 men, at annas 6 per head	4	8	0
Cutting and steeping, 24 men, at annas 6 per head	9	0	0
Stripping and washing at Re. 1-8 per maund, taking 15 maunds as the outturn ...	22	8	0
Rent	6	0	0
TOTAL ...	65	4	0

Profit

Jute of good quality was sold at Rs. 8 per maund (82 lbs.). Thus the price of 15 maunds of jute at Rs. 8 per maund comes to Rs. 120. Net profit, after deducting the costs of Rs. 65 for the cultivation, was calculated at Rs. 55 per acre, or roughly Rs. 50.

*One lakh = 100,000. One standard maund = approximately 82½ lbs.

CHAPTER XIII

Races cultivated on the Burdwan Farm

DURING the harvesting seasons of 1901 and 1902, the seeds of the races of jute cultivated in Bengal were collected by the officers of the Agricultural Department, for experimental purposes, on the Burdwan Farm. When they were grown on the Farm, it was found that the plants were embarrassingly mixed, as usual. The author, who was in charge of the experiments, had repeatedly to eliminate the undesirable plants in order to isolate typical races, after close observation on the experimental fields. They were as follows :—

Species—*Corchorus capsularis*

Nos. 1 to 11.

No. 1.

Race.—*Baran,* from Jagannathganj, Mymensingh.

Description.—The stem and petioles are light green in colour. On 23rd August, 1902, about 2 per cent. of the plants were found flowering. It is the latest of all the eleven races of *Corchorus capsularis* cultivated on the Farm. Not much branchy. The length of the highest plant was 12′-0″.

No. 2.

Race.—*Barapat,* from Jagannathganj, Mymensingh.

Description.—Stem and petioles are light green. On the 20th August, 30 per cent. of the plants were found flowering. Not much branchy. The length of the highest plant was 11′-5″.

No. 3.

Race.—*Kakya Bombai,* from Serajgunj, Pabna.

Description.—Stem and petioles are light green. On 26th August, 10 per cent. of the plants were found flowering. Not much branchy. The length of the highest plant was 10′-9″.

No. 4.

Race.—*Desal,* from Serajgunj, Pabna.

Description.—*Desal* was found to be a mixture of two races of plants. One race has the stem and leaf-stalk of light green colour, while the other has the stem of light green colour, but the colour of the petiole is purplish. On the 20th August, 10 per cent. of the plants were found flowering. Not much branchy. The length of the highest plant was 11′-0″.

No. 5.

Race.—*Barapat,* from Serajgunj, Pabna.

Description.—The stem and petioles are of light green colour, similar to No. 3. On the 20th August, 30 per cent. of the plants were found flowering. The length of the highest plant was 10′-8″.

No. 6.

Race.—*Nailta,* from Serajgunj, Pabna.

Description.—Stem and leaf-stalk are purple. On the 20th August, 10 per cent. of the plants were found flowering. It is strong and not much branchy. The length of the highest plant was 9′-8″.

No. 7.

Race.—*Nal pat* or *Meghnal,* from Rajbari, Faridpur.

Description.—Stem and petioles are purple in colour. On the 20th August, 20 per cent. of the plants were found flowering. It is similar to No. 6. The length of the highest plant was 9′-8″.

No. 8.

Race.—*Amonia,* from Pangsa, Faridpur.

Description.—Stems are light green when young, afterwards purplish, and petioles purplish. On the 20th August, 2 per cent. of the plants were found in flower. A late race ; very strong and branchy. It appears that it is different from *Amonia* (*Amunia*) described in the notes under Races of Jute in Faridpur (No. 12). The length of the highest plant was 9′-7″.

No. 9.

Race.—*Kamarjani,* from Pangsa, Faridpur.

Description.—Stems are light green at first, but purplish when older. Petioles are purplish. On the 20th August, 1 per cent. of the plants was found in flower. A late race ; not much branchy. It is similar to No. 8. The length of the highest plant was 10′-0″.

No. 10.

Race.—*Dhaleswar*, from Narsingdi, Dacca.

Description.—Stem and leaf-stalk are light green. On the 20th August, 1 per cent. of the plants was found in flower. The length of the highest plant was 10′-9″. It is similar to No. 3.

No. 11.

Race.—*Belgachi*, from Madaripur, Faridpur.

Description.—There is a tinge of purple colour in the stems and petioles. A late and strong race, branching freely at the top. The length of the highest plant was 9′-6″.

Species—*Corchorus olitorius.*

Nos. 12 to 15.

No. 12.

Race.—*Tosha*, from Serajgunj, Pabna.

Description.—Stem and petioles of the plants are more or less pink in colour. On the 19th September, 20 per cent. of the plants were found in flower, producing only a few branches at the top. A late race. The length of the highest plant was 11′-0″.

No. 13.

Race.—*Satnala*, from Rajbari, Faridpur.

Description.—The stem and petioles of the plants are light green. It came into flower about 20th July. An early variety, branching irregularly at the top. The length of the highest plant was 11′-0″.

No. 14.

Race.—*Desi pat, Desi lalpat*, from Serampore, Hoogly.

Description.—Stem and petioles are pink in colour. On the 19th September, 26 per cent. of the plants were in flower. The length of the highest plant was 12′-0″.

No. 15.

Race.—*Bangi*, from Narsingdi, Dacca.

Description.—Stem and petioles are light green. The length of the highest plant was 10′-7″. It is similar to No. 13.

CHAPTER XIV

A List of the Names of Races of Jute

NAME.	District where grown.	NAME.	District where grown.
C. capsularis.		*C. capsularis.*	
Lankagoira pat	Hoogly.	Dhupri ...	Faridpur,
Lankagoira pat	24 Parganas.	Ghagri ...	,,
Aus ...	Mymensingh.	Kajla ...	,,
Baran ...	,,	Kamarjani ...	,,
Bara pat ...	,,	Kuchmardan	,,
Boran ...	,,	Lali ...	,,
Chhoran ...	,,	Meghnal ...	,,
Chhota pat ...	,,	Nal pat ...	,,
Kamarjani ...	,,	Peti ...	,,
Kuchmardan	,,	Ranga pat ...	,,
Nal pat ...	,,	Sut pat ...	,,
Parbatya ...	,,	Shani ...	,,
Agniswar ...	Dacca.	Sodu ...	,,
Belgachi ...	,,	Udhap pat ...	,,
Bidyasundar	,,	Bagi ...	Backerganj.
Desal ...	,,	Bangi ...	,,
Desi ...	,,	Cowpla ...	,,
Desi Nallya ...	,,	Nileta ...	,,
Dhaleswari ...	,,	Pechi ...	,,
Dhalsundar ...	,,	Rakta Bagi ...	,,
Kajla ...	,,	Sut pat ...	,,
Parbatya ...	,,	Bidya sundar	Tippera.
Amonia ...	Faridpur	Deo dhali ...	,,
Belgachi ...	,,	Dhal sundar	,,
Cowpla ...	,,	Meghnal ...	,,
Desal ...	,,	Phuleswari ...	,,
Dhajeswari ...	,,	Bara pata ...	Pabna.
Dhapa ...	,,	Bombai ...	,,

NAME.	District where grown.	NAME.	District where grown.
C. capsularis.		*C. capsularis.*	
Deswal ...	Pabna	Hewta ...	Rangpur.
Kakya ...	,,	Hewti ...	,,
Kakya Bombai	,,	Betre ...	Jalpaiguri.
Nalita ...	,,	Bhadoya ...	,,
Ausa ...	Rangpur.	Hewti ...	,,
Bhadoya ...	,,	Lalpata ...	,,
Corchorus olitorius.			
Desi pat ...	Hoogly.	Bami ...	Faridpur.
Desi lal pat ...	,,	Bara pata ...	,,
Baran ...	Mymensingh.	Dhala Bagi ...	,,
Bar pata ...	,,	Dhamraj ...	,,
Nileta ...	,,	Meghnal ...	,,
Packnallya ...	,,	Nal Bagi ...	,,
Talla ...	,,	Ranga Bagi ...	,,
Banghi ...	Dacca.	Satnala ...	,,
Banghi nallya	,,	Parbatimadai	Rangpur.
Deo nallya ...	,,	Bomochak ...	Jalpaiguri.
Bagi ...	Faridpur.	Marua pata ...	,,

PART IV—TRADE

CHAPTER I

Physical Divisions of Jute Tracts

SCIENTIFICALLY the jute areas may be divided into two great divisions according to the courses of the two great rivers, the Ganges and the Brahmaputra, which, with their tributaries, flow through the jute areas. The characters of the soils of the Gangetic area on which jute is grown and the waters of this area in which the jute is steeped and washed—the two principal factors for producing the crop—are quite different from those of the Brahmaputra area. The lands in the Gangetic area are for the most part low, subjected to annual floods, while those in the Brahmaputra area are higher. The water of the Ganges is muddy, while that of the Brahmaputra is clear. These differences in the conditions of the soils and the waters of the jute areas give distinct effects to the qualities of the fibre. An expert trader in jute will separate the fibre of one division from that of the other without any geographical knowledge of the tracts where they are produced. The fibre of the Gangetic area is strong, coarse, barky and yellowish in colour, which fades in storage, while that of the Brahmaputra area is fine, white, more glossy and elastic, as well as strong. The fibre of the Brahmaputra area can bear more tension and twisting. Of course, there are exceptions. The colour of the jute of the *bheel* lands in the Gangetic area, where there is clear water for washing, is in no way inferior to that of the good-coloured jute of the Brahmaputra area. On the other hand, in parts of the Brahmaputra area where clear water is not available, steeping and washing is done in dirty ditch water, and the colour of this jute is not better than that of the jute of the Gangetic area.

There are high lands containing loamy soils in the Gangetic area also where jute of fine texture is obtained. On the other hand, coarse fibre of fading colour is produced

on the low *char*-lands in the Brahmaputra area. Barring these exceptions, however, the fibres of these two divisions may be distinguished on the distinct points noted below :—

Fibres of the Gangetic Area	Fibres of the Brahmaputra Area
1. Strong.	1. Strong.
2. Coarse or thick.	2. Fine.
3. Short.	3. Long.
4. Yellowish coloured.	4. White coloured.
5. Barky (Specky).	5. Little barky.
6. Less glossy.	6. Glossy.
	7. Bearing more tension and twisting.

In a word, the fibres possessing all the qualities which are desirable are available from the Brahmaputra area.

Jute grows to a very small extent (20,000 acres in 1920, and 21,000 acres in 1930) in some parts of the delta formed by the rivers Mahanadi, Brahmani and Baitarani included in the districts of Cuttack and Balasore in Orissa. The quality of the fibre of this part of the country is coarse and yellowish, and also sticky, due to defective stripping.

In the Gangetic area, the jute crop is first seen in the *diarah* (*char*) lands in the Rajmahal Subdivision in the district of the Sonthal Parganas on the right bank of the river Ganges. The Ganges has taken a slower course down from Rajmahal. The cultivation has not extended, to any extent, on the high lands in the Gangetic plains known as the old alluvium consisting of hard reddish soils.

From Rajmahal down to the Hoogly or Bhagirathi, the western branch of the Gangetic delta, the cultivation is entirely confined to the *diarah* lands. But it is extensively cultivated on the higher parts of the delta which commences about 300 miles off from the sea, excluding salt-impregnated parts in the Sundarbans. This area includes the districts of Hoogly and Murshidabad on the west, Nadia, the 24-Parganas, Jessore and Khoolna in the middle, and Faridpur and Barisal on the east. The most important branches of the delta are the Bhagirathi or Hoogly on the west, the Bhairab and the Goroi or Madhumathi in the middle, and the Arialkhan on the east. Most parts of this area are annually inundated, where silts are deposited. Special characteristics of the fibres of this tract are coarse in texture and yellowish in colour, fading in storage. The *daisi* jute is principally obtained from the banks of the Hoogly.

On the left bank of the Ganges, the jute area has, properly speaking, commenced downwards from the Kushi river, which divides Purnea from Bhagalpur. Jute is cultivated to a very limited extent further up in the Tirhut Division in North Bihar. This tract of land is watered by a net-work of tributaries of the Ganges, of which the most important are the Gogra, the Gandak and the Kushi. They overflood the banks suddenly to a depth of water which the crop cannot resist. Otherwise, the soil which is rich and alluvial is well suited to it. After leaving the branch, Bhagirathi (Hoogly), on the right, the Ganges takes the name of Padma. From the Kushi on the west to the alluvial banks of the Brahmaputra on the east, jute is very extensively cultivated. This area may be divided into three parts or divisions. The first part consists of a tract of low lands between the Kushi on one side and the Mahananda on the other. The latter flows down slowly to the Ganges through the districts of Purnea, Dinajpur and Maldah. This part of the country may be included in the Gangetic area. Fibre of an inferior quality, in points of strength, length and colour, is produced here. The weakness of the fibre is due to over-steeping and not to the cultivation of an inferior race. The second part of this country consists of the high lands between the Mahananda and the Karotoa rivers, including parts of the districts of Dinajpur, Jalpaiguri, Rangpur, Bogra, Rajshahi and Pabna, with the exception of a contiguous and large tract containing old alluvial soils of a structure similar to that found on the right bank of the Ganges. This section, called *khiar* land, contains portions of the districts of Dinajpur, Rangpur, Bogra, Rajshahi and Maldah. The word *khiar* is most likely derived from the colour, which resembles that of *khair* (*kath*). The Karotoa flows in a slow course towards south-east, from Darjeeling, to the main stream of the Brahmaputra called Jamuna. Another important river in this part is the Atrai. It flows through the districts of Dinajpur and Rajshahi. These two districts are almost bisected by it. The banks are rarely overflooded. It then joins the Karotoa river in the district of Pabna. The Northern class of jute comes from this part of the country. It is tall, fine, and sound, and can bear twisting very well, but is generally yellowish in colour owing to its being steeped and washed in ditch water, and it is somewhat weak for over-steeping. Otherwise this area yields a fibre of a high grade. The Brahmaputra area commences from the river Mahananda. Now comes the third part of these jute

tracts lying between the river Karotoa, a tributary to the Brahmaputra, and its main course. This part of the country is watered by the Brahmaputra and its other numerous tributaries, of which the important ones are the Subansiri and the Manes draining, the extreme northern valley of Assam, and also the Gadadhar or Sankosh and the Raidak. The tributary Raidak separates Assam from Bengal. All of them flow direct from north to south. The most important of the tributaries in this division of the Brahmaputra area is the Teesta. It rises from the Himalaya and takes a south-eastern course from Darjeeling. In its course to the Brahmaputra, it drains parts of the districts of Jalpaiguri, Cooch Behar and Rangpur. The most important river on the eastern part of the Brahmaputra area is the Surma, which flows through the districts of Cachar and Sylhet into the Meghna. They form two valleys in Assam, called the Brahmaputra and the Surma Valleys. Jute is grown to a limited extent in the Surma Valley, which does not appear to be well suited to the crop, owing to heavy rainfalls from the beginning of the growing season. The Brahmaputra Valley, on the other hand, is better adapted to the crop, where it is steadily extending. The fibre of the Brahmaputra Valley and that of the banks of the Teesta are identical and of good quality, and should form a separate class in the trade.

The Brahmaputra, after breaking through the mountainous regions in north-eastern Assam, takes a slower course towards the west, watering the plain lands of northern Assam, called the Brahmaputra Valley, within the limits of the districts of Lakhimpur, Sibsagar, Nowgong, Darrang, Kamrup, Goalpara and the Garo Hills. On leaving the Garo Hills on the left, it turns abruptly towards the south and enters Bengal. After it receives the Teesta near Fulchari in Bogra, it flows down in a rapid course to join the Ganges at Goalundo, overflooding parts of the districts of Rangpur, Bogra and Pabna on the right bank and Mymensingh and Dacca on the left. The jute fibre of this section of the Brahmaputra or Jamuna is not very good in quality, being rather coarse and yellowish in colour, and fades in storage. In other respects it is a long, strong and healthy class of jute.

The Brahmaputra, in another very slow course, flows down through the Mymensingh district, from the north-western corner near Bahadurabad to the south-eastern corner, and joins the Meghna. This course is now called the old Brahmaputra. Formerly the main current flowed

through it. The present main stream, already noticed, which meets the Ganges at Goalundo, is still called by the name of Jamuna by the people who live on its banks. The old Brahmaputra and the Meghna supply the country with exceptionally clear water. The highest class of jute is produced on the high lands of the banks of the old Brahmaputra and Meghna. No other quality can approach it, in point of excellence, for which fancy prices are always offered.

The Barak (Barabakra) which rises from the Lushai Hills joins the Surma near Marculi. It is the most important tributary of the Surma. Many other tributaries flowing through the eastern parts of Mymensingh and western parts of Sylhet, fall into the Surma on the right bank. These parts also form a most important jute tract. The Surma from Marculi downwards is called by the name of Meghna. According to some eminent writers on the geography of India, the Surma is said to be a tributary to the Brahmaputra. But from the fact that the Brahmaputra joins the Meghna at almost right angles, it may be suggested that the Surma or Meghna is not a tributary, but is probably older in origin than the Brahmaputra. It is not our point, however, to discuss the matter further as to the ages of these rivers.

From the junction of the rivers Brahmaputra and the Meghna, it flows in a south-western direction, leaving the districts of Sylhet and Tippera on the left bank and Mymensingh and Dacca districts on the right, and joins the Ganges over again near Chandpur, about 70 miles off from the Bay. They retain the name of the Meghna till they fall into the sea. A high class of jute is obtained from the banks of the Meghna up to the Chandpur Station. From Chandpur downwards the crop suddenly disappears.

Thus we see that jute is a romantic plant. It thrives in particular soils under certain climatic conditions, but avoids some soils of the neighbouring places under the same climatic conditions. It becomes fine in some soils, but coarse in others, at the same place. The fibre changes its colour for slight differences in waters used for steeping and washing. It has adapted the soils of the deltas of the Mahanadi and the Ganges and the soils of the banks of the Brahmaputra and the Meghna, but quietly rejected other parts of the country. It has revolutionised the Indian trade, and the people who deal in any way in it are getting prosperous.

Taking physical conditions into consideration, the areas under jute, producing different qualities of fibres, may be put into the following divisions :—

(A) Gangetic Area

(1)	(2)	(3)	(4)
Madaripore	Jessore	Hoogly	Purnea

(B) Brahmaputra Area

(1)	(2)	(3)	(4)	(5)
Upper Brahmaputra	Lower Brahmaputra (Jamuna)	Old Brahmaputra	Meghna	Northern

In the Gangetic area, the Hoogly division yields the *Daisi* quality.

The Old Brahmaputra division supplies the commercial quality of *Jath* jute. The *District* quality is principally imported from the Upper Brahmaputra, Lower Brahmaputra and Meghna divisions.

CHAPTER II

Commercial Divisions of Jute Tracts

JUTE growing tracts in Bengal were commercially divided into five Divisions in the first edition of this book, in 1908, *viz.* :—

(i) Narainganj ;
(ii) Serajganj ;
(iii) Uttarya or Northern ;
(iv) Dowrah ;
(v) Daisi.

(i) Narainganj jute is grown on the tracts of lands which are supplied with water from the old Brahmaputra river. There is no other jute tract in Bengal where water is so clean as that of the old Brahmaputra. The colour of the jute is the best in the market. Most parts are however liable to submersion by floods when the crop is still on the fields. This standing water is responsible for the *specky* and *mossy* fibres of this tract. High lands yield fibres of exceptionally good quality. About 30 per cent. of the jute of this tract is "hessian." Narainganj and Chandpur are the principal markets which are under the jurisdiction of Mymensingh, Dacca and Tippera districts.

(ii) Serajganj jute is obtained from the tract of lands which are fed by the new Brahmaputra or Jamuna river. The water of this river is not so clean as that of the old Brahmaputra. The principal market of this tract is Serajganj, which is supplied with jute from the western parts of Mymensingh, and from Pabna, Bogra, Cooch Behar, Rangpur and Goalpara. About 30 per cent. of the jute is "hessian."

(iii) Uttarya or Northern jute is grown in the tract of high lands which are principally watered by the tributary rivers of the Brahmaputra. Jute of this tract is steeped and washed mainly in ditch water, and as the same water is used for successive steepings, it becomes more or less coloured owing to decayed vegetable substances. The colour of the jute is therefore inferior. Thirty per cent. of this jute however comes under "hessian." Huldibari,

Domar, Kissenganj, Kasba and Forbesganj are the principal marts in this tract. This area includes the districts of Rajshahi, Bogra, Rangpur, Jalpaiguri, Dinajpur, Maldah, and Purnea.

(iv) Dowrah jute is grown on the lands which are swamped by the branches of the river Ganges, containing muddy water. This muddy water imparts a grey colour to the fibre of this tract. The *Dowrah* jute is very strong, but specky, mossy and harsh. It is chiefly used for sackings and cordage. Madaripur, Berhamganj, Angaria and Gopalganj, in the district of Faridpur, are the chief markets in this tract.

(v) Daisi jute belongs to the species *Corchorus olitorius*. It is grown on the high lands in the neighbourhood of Calcutta. The jute of this tract is steeped in the ditch water supplied by the rains or the rivers Bhagirathi and Damoodar, also Roopnarain, which contain muddy water during the rains. Owing to the quality of the water and some objectionable method of steeping, the jute of this tract is more or less black in colour. This jute is principally used for bagging. Badyabati in Hoogly and Belgatchia in the 24-Parganas are the important markets.

The above classifications were made by the trade according to the different qualities of the fibres imported into the markets of a Division. But the fibres of a Division were certainly not of uniform qualities. Now, for the convenience of the trade, systematic classifications of the jute markets have been made more or less according to the qualities of fibres sold. Thus in the same tracts, markets of different classes are found. For example, in the old Naraingani tracts, comprising parts of Mymensingh, Dacca and Tippera districts, there are two classes of markets—one class selling the best grades of fibre called "Jath" (true) jute, while in the other the fibres of inferior qualities known as "District" jute are sold; but they are only a few in number.

"Jath" jute is fine, strong, long, healthy and glossy, of milky white colour, and "District" jute is coarse and more or less specky or mossy, the colour of which fades afterwards and the healthiness is lost. In other respects, it is similar to the "Jath" *pat*.

The "Jath" and the "District" jutes are grown on the same tracts, although there are differences in qualities due to the facts that "Jath" jute is the product of the high lands, while "District" jute is the fibre of the *char*

lands of the same Brahmaputra area. In some "District" jute markets, the fibres, both from the Gangetic and the Brahmaputra areas, are sold, which cannot however be taken as general. "District" jute consists entirely of the fibre of the *char* lands of the Brahmaputra area.

According to the present commercial classifications, the "Northern" jute does not fall either under "Jath" or "District" owing to the facts that it is blackish and weak on account of steeping and washing in dirty ditch water and over-retting, by which it loses its gloss and healthiness. Otherwise, it should have been a high-class jute of fine texture.

The jute of the defunct Dowrah Division now comes under Madaripur quality. The fibres of the Gangetic area already described fall under this class, which is inferior to the fibres of the qualities "Jath" and "District." This class of fibre of the Madaripur Division is coarse and more or less specky and mossy, though strong. The colour is yellowish, which fades in storage. The jute which is produced on the left side of the river Goroi or Madhumati, goes to the Madaripur quality, while that on the right is assorted as the "Jessore" quality. It is similar to that of the Madaripur class, but it is not so strong, owing to over-steeping. The Purnea jute is identical with that of the Jessore class, but it is generally shorter. A better class of fibre is imported into the Purnea markets from Nepal.

The Mahanadi Division in Orissa yields a strong and yellowish fibre, but it is more or less sticky owing to defects in stripping. In this Division jute is chiefly cultivated in the Kendrapara Subdivision in the Cuttack district.

There has not been any change in the *Daisee* quality of jute. It belongs to the species *Corchorus olitorius*. *Tosha* of Serajganj, *Bagi* of Dacca or Faridpur, and *Nileta* of Mymensingh come under the same class, but of superior quality. *Daisee* jute is largely cultivated in the Burdwan Division and in the 24-Parganas in the Presidency Division in Bengal. In Eastern Bengal and Pabna, its cultivation is but limited, not exceeding 5 per cent. of the total crop, while it is little grown in Northern Bengal. It is never cultivated in the low lands subjected to floods. The fibre of this quality is coarse, brownish yellow in colour, a little specky or mossy, and healthy. It is more suitable for sackings than for hessian. Owing to the dirty water of ditches in which this jute is steeped in the neighbouring districts of Calcutta, the colour of the fibre becomes blackish.

The following diagram shows the various qualities of jute which are available in the markets for sale :—

Divisions of Qualities

I	II	III
Gangetic Area	Brahmaputra Area	Mahanadi Area

I. Gangetic Area

(1)	(2)	(3)	(4)
Madaripur	Jessore	Purnea	Daisee of West Bengal Bangi of East Bengal

II. Brahmaputra Area

(1)	(2)	(3)	(4)
Jath	District	Northern	Tosha Bagi or Nileta of East Bengal

III. Mahanadi Area

(1)
Cuttack

It should be noted that these classifications have nothing to do with assortments of fibres in which business is done in the market. The fibres of any particular class, such as Jath, District, Northern or Madaripur, are again assorted as 1's, 2's, 3's, 4's, etc., according to the qualities of the fibres contained in each bale as enumerated in Chapter V, Part IV. But the prices of the fibres of these assortments differ according to the classes they belong to. Thus the fibre of the assortment of 2's or 3's of the Jath class of jute is sold at a higher price than that of the assortment of 2's or 3's of the District or the Madaripur or the Northern jute. Similarly, the 2's or 3's of the District class of jute are considered to be better than the 2's or 3's of the Madaripur class. In the same way, differences are made between 2's or 3's of *Tosha* (*C. olitorius* of North Bengal) and *Daisee* (*C. olitorius* of West Bengal). A list showing the markets dealing in different classes of jute, as classified above, is given elsewhere.

CHAPTER III

Classification of Fibres

JUTE fibres are assorted in the trade according to their differences in qualities, such as strength, length, colour, gloss and fineness. In fine or soft jute there is absence of root, bark or specks and gummy substances; while harsh jute may contain one or more of these defects. Weak or damaged jute loses its strength and also its gloss. There are six recognised classes of jute, as follows :—

I. Hessian-warp ;
II. Hessian-weft ;
III. Sacking-warp ;
IV. Sacking-weft ;
V. Rejections ;
VI. Cuttings.

I. Hessian-warp is a strong, long, healthy and fine jute of good gloss and silvery white colour, without any defects, such as barky, mossy, sticky or specky. It is used for spinning thread, which is suitable as warp for weaving hessian cloth or other finer fabrics. This class of jute consists of the finest fibre of the country. Only a limited quantity of this quality is available. It is chiefly exported to Europe.

II. Hessian-weft is inferior to Hessian-warp in length and gloss, but similar to it in other respects. This jute is used for weft in the hessian cloth referred to above.

III. Sacking-warp is a strong, long and healthy jute, irrespective of any particular colour. It is coarser than I and II. The thread spun out of this jute is used as warp for weaving sacks.

IV. Sacking-weft is a dull-coloured short* or weak jute used for weft in sacks.

* The fibre which is below 6 feet in length, after deducting the cuttings, is called "short" jute, the usual length of long jute being from 7 to 12 feet. In a normal year, two-thirds of jute are long and one-third short.

V. Rejections are the worst stuff. They are either damaged, barky or knotty, and used in spinning thread for sacking weft. Barky jute is used in spinning after it is soaked with oil and water and passed through softeners.

VI. Cuttings are the barky or hard portions of jute cut off from the bottom or the top. Cuttings are either used in spinning sacks as weft, after being softened with oil and water as referred to above, or in paper making.

CHAPTER IV

Jute Presses

JUTE is pressed into bales by the dealers for convenience of transport from one place to another. The baling is done by means of hand and steam power presses. The bales which are pressed by hand press are called *katcha* bales, while those done by steam press are *pacca* bales. The *katcha* bales are re-baled into *pacca* bales by the balers, if they are intended for export out of India. A *pacca* bale contains fibres of more uniform quality according to definite standard of a *mark* of a baler.

A large number of hand power presses are used in this country for pressing bales weighing 3½ maunds (280 lbs.) each, not exceeding 13½ cubic feet in measurement. There are bales of 4 maunds (320 lbs.) each also, sized as 4′-0″ ×2′-0″ ×1′-6″ by steam press to a density not exceeding 12½ cubic feet. In some places, where facilities for conveyance are not sufficient, smaller presses for bales weighing 1½ maunds (120 lbs.) each, sized to 2′-9″ × 1′-8″ × 1′-3″=5·8 cubic feet, are found. From an economic point of view, the former are preferred to the smaller bales.

There are two classes of presses sold in Bengal for the *katcha* bales: one class is called the lever press, and the other the hydraulic press. Lever presses are simpler in construction, and it is easy to fit or set them up. It requires some mechanical training on the part of the man who would fit up a hydraulic press. The hydraulic presses can be economically worked with steam power, in which case one set of pumps, working two presses at a time, gives much more outturn than that obtained from two hand-power lever presses. European buyers have been introducing steam power for pressing in mofussil markets.

Ordinarily one lever press, working 10 hours a day, turns out about 200 bales of 3½ maunds each with 16 labourers, or about 200 bales of 1½ maunds each with 8 labourers, at the rate of 20 bales per hour. The act of pressing is commonly done by contract system. For each bale of 3½ maunds, about two annas, while for 1½-maund bales, about an anna-and-a-half per bale, are paid to the

labourers. The total cost, including the hire of the press, godown, labourers, etc., amounts to 8 annas per $3\frac{1}{2}$-maund bale and 6 annas for each $1\frac{1}{2}$-maund bale. By steam power, the cost may be reduced by 2 annas per bale.

Hand-presses are sold by many firms in Calcutta. The Chain Press of Messrs. Mandal & Co. is claimed to be cheaper than any other lever presses.

For export, out of India, bales of 400 lbs. (5 maunds) each are pressed to a density of 10·4 cubic feet, being reduced now to 9·44 cubic feet, thus making a ton of 50 cubic feet in density, by five bales.

There are three patterns of pressing machines in Calcutta, the best of which is called the Cyclone Press. It has a capacity of pressing about 60 bales of jute or 40 bales of cuttings of jute per hour. The Watson Press comes next to the Cyclone Press. About 40 bales of jute or 26 bales of cuttings may be pressed in an hour by this machine. The capacity of pressing of the Nasmyth Press is about 30 bales of jute or 20 bales of cuttings per hour. The most modern press produces 60 to 80 bales per hour. Rs. 2-8 is charged for pressing a bale of jute, including the costs for cordage (5 lbs.) and despatch from the Press-house. For a bale of the cuttings or *Daisi* jute, annas 2 extra is charged for a pound of additional rope required per bale. An additional amount of about annas 14 for the jute of the species *C. capsularis,* annas 10 for *Tosha,* and annas 5 for *Daisi* (which does not require the roots to be cut off) per bale, is charged for assortment and cutting the roots.

For *pacca* bales, there are 25 pressing machines in Calcutta, 2 at Narainganj and 1 at Chandpur. Each machine has a capacity of working 3 or 4 presses. Altogether there were 110 steam presses working in Bengal during 1928. About 30 men are required for driving a machine for *pacca* bales—*mistries,* firemen and *khalasies,*—and an additional 20 men for each press for baling. A vast army of labourers are wanted for assortment. One man can sort fibre for $1\frac{1}{2}$ to 2 bales of *C. capsularis* (*Bilathi pat*) and 4 to 5 bales of *Daisi* jute, per diem.

CHAPTER V

Loose Jute

JUTE is exported to the Calcutta Bazar, the mills and the press-houses, in drums or *katcha* bales. The drums contain 30 to 40 seers of jute each, without any particular assortment. The *katcha* bales are generally of 3½ maunds (280 lbs.) each, assorted to a certain extent, and pressed at the places from where they are despatched. The system of grading has been considerably changed, since the second edition of this book was published. It was as follows :—

1's containing ... 80 to 90% Hessian-warp ;
10 to 20% Weft and cuttings.

This quality was entirely exported to Europe, when available.

2's ... 40 to 60% Hessian-warp ;
40 to 60% Weft and cuttings.

2's divided into the following groups—

2's (Green) ... 60% Hessian-warp ;
20% Sacking-warp ;
20% Weft and cuttings.

Calcutta mills used a very small quantity of this quality. It was mainly requisitioned for Europe.

2's (Red) ... 50% Hessian-warp ;
40% Sacking-warp ;
10% Weft and cuttings.

2's (Ordinary) 40% Hessian-warp ;
40% Sacking-warp ;
20% Weft and cuttings.

Red 2's were largely used by the Calcutta mills. From this stuff, the mills picked up the fibre suitable for weaving hessian cloth.

There was another assortment under 2's called X2's which also had a large business in Calcutta.

X2's ... 20% Hessian-warp ;
60% Sacking-warp ;
20% Weft and cuttings.

3's ... 70% Sacking-warp ;
30% Weft and cuttings.

Good 3's ... 10% Hessian-warp ;
60% Sacking-warp ;
30% Weft and cuttings.

Under 3's there was also another quality which was called in the Calcutta market as X3's.

X3's yielded ... 60% Sacking-warp ;
40% Weft and cuttings.
4's ... 40% Sacking-warp ;
60% Weft and cuttings.
Good 4's ... 50% Sacking-warp ;
50% Weft and cuttings.
X4's ... 30% Sacking-warp ;
70% Weft and cuttings.

There were too many qualities in which business was done in the Calcutta market. The result was that there were frequent disputes between the buyers and the sellers.

Rejections contained all weft for sacking. It was the worst of all the fibres sold. But a better quality of rejections yielded :—

20% Sacking-warp ;
80% Weft and cuttings.

A very inferior grade of rejections was also available in the market, which contained fibre which was barky, specky and hard all through, due to over-ripening of the crop and bad steeping and washing. It was called "Terrible Rejections" (T. R.), and used chiefly for paper manufacture.

Cuttings were the same as found at present. They contain portions of jute cut from the bottom or top of the fibre to bring it under a uniform quality. Cuttings are of different qualities as noted below :—

N. C. = Narainganj jute cuttings.
O. C. = Ordinary jute cuttings.
T. C. = Tosha jute cuttings.
D. C. = Daisee jute cuttings.
B. C. = Bagging cuttings.
M. C. = Mixing cuttings containing all sorts of cuttings.

Cuttings, in mixture of short jute, are used in the Calcutta mills in the manufacture of sacking cloth as weft, in the following proportions :—

2 parts short jute,
1 part cuttings.

Narainganj cuttings, which are 10 to 12 inches long, are the best of all. Bagging cuttings are used for paper making in Europe. They are very short, about 4" long, as they are obtained from the cuttings of bags. Mixing cuttings are the worst stuff, and cannot be used for baggings, but for paper only.

Cuttings are largely used in the Indian mills for mixing purposes; they are also baled for export to foreign countries.

The standard grade of the loose jute market is called 50/50, that is, 50 per cent. of the bales contain 2's and 50 per cent. 3's. The price of this standard grade is the basis on which the prices of other qualities are determined. Of course the price of a particular quality chiefly depends on its demand.

We had much to say about the foregoing classification of the loose jute market. There were some confusing marks, such as red 2's, green 2's, ordinary 2's, X3's, X4's, etc. The Indian Jute Mills Association has, however, taken up the subject of classification in right earnest.

We are sure there would be much difficulty in changing the trade methods of classification. First of all, the sellers would not be willing to call a 2's of the market as the 3's of the proposed revised grades.

Although there may be practical difficulties in the re-arrangement of groups, it will undoubtedly simplify the present complications existing in the trade. On several occasions, we had to classify the fibres of different centres. For each quality, certain marks were allotted, and the marks were distributed to each sample for each quality. The sample securing the highest marks was placed at the top. This system worked satisfactorily, and the European jute merchants, who were invited in 1903 and 1904 to make a valuation of the different samples of jute of different centres, approved of this system of examination, and admitted that the examination made in this way was more or less accurate.

The marks were allotted as follows :—

Quality.	Full marks.
(1) Strength	20
(2) Length	20
(3) Colour	20
(4) Absence of roots and other defects ...	30
(5) Fineness	10
TOTAL FULL MARKS ...	100*

Strength.—The fibre should not break easily when pulled or twisted. When it breaks it gives a hissing sound.

* The fibres are supposed to be free from fraudulent watering. The dry jute of commerce contains 8 to 10 per cent. of moisture, which is accepted as natural moisture in jute fibres, during the rainy season.

Standard length of fibre for full marks may be 10 to 12 feet with roots, not below 9 feet after the roots are cut. The top qualities should have a creamy white colour. Well grown fibre free from disease is said to be sound. Such defects as specky, knotty, croppy, *ghata*; etc., should be absent from sound jute. Fineness and absence of roots require no explanation. It may however be said that from 6 to 18 inches are commonly cut from the bottom of fibres for spinning. Marks would be awarded according to the length of fibre thus left after rejecting the "cuttings."

It is hoped that the difficulties of grouping of fibres, both of the loose and baled jute markets, would thus be greatly minimised.

The Indian Jute Mills Association is trying to revise the assortments of marks, as a temporary measure, as follows :—

The marks of the crop are to be as follows: —

1's,
2's,
3's,
4's and
R's.

The mark 1's to represent jute of superior quality unsuitable for general mill consumption in India.

The mark 4's(a) to be straight *mora* head and to contain not less than 40% to 50% of sacking warp, and 4's(b) to be sold on the standard of the mark of mills' requirements.

R is to be the lowest mark, and the arrangement regarding 10% *hubby-jubby* (mixed jute) to be abolished.

The marks 2's and 3's are to be sold on the standard of the mark, or on guarantee terms, according to mills' desire.

Weedings to be packed and sold separately as usual.

On account of a complication between the mills and the sellers which arose in 1930, it was necessary to introduce a mark L. R. which is lower than R.

The proposal of arranging samples representing various marks, which should be obtained and exhibited in the Arbitration Room of the Association with a view to prevent any possible misunderstanding between the parties, was not accepted. This was a proposal of great consequence for the sake of smooth dealing in the jute business in which frequent disputes arise between the sellers and buyers regarding qualities. It is hoped that the Association will introduce this reformation in the near future.

CHAPTER VI

Baled Jute

JUTE is exported from Bengal to Europe and America, in bales of 400 lbs. each, pressed by steam power to the measure not exceeding 10½ cubic feet, 4′-0″ × 1′-9″ × 1′-6″ in size, being reduced now to 9·44 cubic feet, to occupy 4′-1″ × 1′-6½″ × 1′-6″ of space, of which 5 bales make precisely a ton of 50 cubic feet. A bale contains definite qualities of fibre, assorted according to definite marks of the balers, who are members of the Calcutta Baled Jute Association.

Each specified mark represents a particular quality and grade of jute duly registered by the Calcutta Baled Jute Association. No public mark that has not been included in the Official Book of Marks issued by the Calcutta Baled Jute Association, on or before the first July, in each year, and lodged with the London and the Dundee Jute Associations, can be substituted under any grade. In the event of unavoidable circumstances preventing shipment of such mark or marks as guaranteed, the seller can substitute another mark or marks of the same grade and of equal quality, at a reduction of 10*s*. per ton, from contract price. If it prove to be inferior to that of guarantee, the liability of the seller to the buyer shall be on the mark or marks specified. No allowance can be claimed unless the inferiority shall exceed 10*s*. per ton, and in that case the buyer shall have the right either (1) to recover from the seller the full difference in value on the date when such inferiority shall be acknowledged or final award issued, or (2) if the inferiority shall exceed 50*s*. per ton, to invoice the jute back to the seller within 5 days after receipt of the award and recover from the seller, in respect thereof, a price per ton to be stated in the award, equal to 10*s*. above the market value of the quality contracted for, on the date when such inferiority is acknowledged or final award issued. On substituted marks, the buyers shall have the right either (1) to recover from the seller the full difference in value; or (2) if the inferiority shall exceed 20*s*. per ton,

to invoice the jute back to the seller within 5 days after receipt of award, and recover from the seller, in respect thereof, a price per ton to be stated in the award, equal to 10*s*. above the market value of the quality contracted for, on the date when such inferiority is acknowledged or final award issued.

The marks are grouped as follows :—

Diamonds	Lightnings
Reds	Mangos
Firsts	Hearts
Daccas	Daisi

Jute is sold under guarantee* in Calcutta, which is to be of the average standard of the grade, free from Bombay quality and Rejections, and the roots being clean cut. This guarantee is restricted to jute of the grades higher than what is known as "Hearts," with an exception of the "Daisi" group. Formerly the standard quality was known in the baled jute market by the name of *M* group or "Cracks," baled principally out of the jute of the Serajganj Division. This group is now substituted by the "Firsts," which are made up of the selected qualities of the Northern jute.

This group, as well as those below it, excepting "Daisi," is composed of the fibres in equal proportions of 2's and 3's, packed, of course, separately. In the "Firsts" 2's should yield 40 per cent. of Hessian-warp and 3's 70 per cent. of Sacking-warp, not without a large percentage of Hessian-weft in the latter. Now true Hessian-warp is hardly used for the marks of this group. This quality of fibre is specially picked up for the marks of the higher group. At present, in the "Firsts" the 2's contain 40 per cent. of warp jute resembling Hessian-warp in all respects, specially in length and strength. Next to the "Firsts" the "Daccas," assorted similarly as the "Firsts," form a group which is most important. They are baled chiefly out of the District qualities of the Mymensingh and Dacca districts. Differences between these two groups are on the following points :—

"Firsts" consist of soft fine jute of inferior colour, while the Daccas are made of hard coarse jute of superior colour. Both are equally long.

---

* For Details, please see the Contract Forms issued by the London Jute Association and the Calcutta Baled Jute Association.

Taking all points into consideration, there are slight differences between them. Sometimes "Daccas" are sold at a little higher price. There is always a strong demand for it.

The marks of the "Lightnings" group are made up of the medium common Northern qualities, while the "Mangos," popularly called the C. D. M. group, out of the common medium District jute. In the marks of both the groups, the residues of the higher marks are used. The fibres of these groups are largely used for sacking purposes.

Formerly, the Madaripur qualities had no place in the baled jute market, due to excessive malpractice of fraudulent watering and also to its containing tap roots (Bombay Quality), as the plants were uprooted instead of being cut. These defects have lately been eliminated. Now the Madaripur jute is eagerly purchased for the "Mangos" marks, also for the "Daccas." Some portion of it, which is produced in the *bheel* tracts, goes still higher for the "Diamonds."

It may be stated that the C. D. M. (red mark) under the "Reds" is quite different from the popular C. D. M. C. D. M. (red) is superior to the marks of the group "Firsts" or the "Daccas."

"Hearts" are composed of low grade weak jute with which gunny bags are chiefly manufactured.

As already noticed, there are some marks of fine and exceptionally good qualities with creamy white colour of true Hessian-warp.

The "Diamonds" are packed out of the picked fibres of the Mymensingh and Dacca districts, while the "Reds" from the "Northern" qualities. Only a limited business is done in the fibres of these two top groups. Dundee is the principal buyer, and also France and Germany, where these fibres are utilised for mixing purposes for manufacture of woollen textiles.

It may be noted that different marks, even under the same group, differ considerably from one another, both in qualities and values.

Besides the foregoing marks, there are also some recognised marks of inferior qualities consisting of "Bombay" or "Dowrah" qualities, "Rejections" and the cuttings of different qualities which were already noticed in the chapter under Loose Jute.

In addition to the groups of marks described above, there is another group known as the "Daisi," composed of the jute of *C. olitorius* species. It is getting into favour with buyers both in Europe and India. The importance of this quality lies in its absence of roots, which is a thing of great economic gain to the spinners. It is sent out to Europe in three principal numbers of 1's, 2's and 3's. The general standard is an equal percentage of 1's and 3's, for example, 10 per cent. of 1's, 80 per cent. of 2's, and 10 per cent. of 3's, or 20 per cent. of 1's, 60 per cent. of 2's, and 20 per cent. of 3's. These may be illustrated by the following equations ;—

1's. 2's. 3's.

10+80+10=100 or

20+60+20=100 or

30+40+30=100

Some of the standard marks under the aforesaid groups recognised by the London and the Dundee Jute Associations are shown overleaf.

The foregoing paragraphs have been written after discussion with some of the prominent balers of Calcutta, and also after consulting their documents, according to which the baled jute business is at present done. From this, we are afraid, none of the readers, unless he is a trader in baled jute, has been enlightened as to the qualities of jute contained in the bales of a particular mark or marks registered by the Calcutta Baled Jute Association and noted in the book "Jute Marks and Assortments." Not even the different existing groups of marks of the bales will help any one to ascertain as to the particular quality or qualities of which a group is composed. The quality of one mark differs widely from that of another mark under the same group. On the other hand, it will appear that from the top to the bottom of the groups, each mark is said to contain "2's" 50 per cent. and "3's" 50 per cent. If so, there should not have been any differences and so many divisions or groups of marks. Thus the 2's or the 3's have no significance at all. The present system of marking is so defective and confusing that even the balers have no control over the assortments made by their subordinates. The buyers too, specially those of foreign markets other than London and Dundee, are put into great difficulties in the work of checking the qualities of the different marks, which are increasing every year. It is not practicable to

have samples of different marks before each member of the buyer's staff through whom the transactions are carried on.

Mr. Antonino Neves, who came to India in 1920, as a representative of the Brazil Republic, to study the subject of jute and its trade of this country, told us that they were simply helpless in the hands of the sellers regarding the qualities of jute supplied to his country. Many of the Calcutta balers with whom we had discussions on the subject, agreed with us in the various main points of defects in the marks of the Calcutta baled jute.

It may, we think, simplify the baled jute trade if the sellers and the buyers would agree to accept the classifications of the qualities adopted in the loose jute market of this country. The system of grading of "Daisi" jute seems alright. The main jute fibres may be graded similarly. The standard quality may have 2's, 3's and 4's in the same ratio as in the "Daisi."

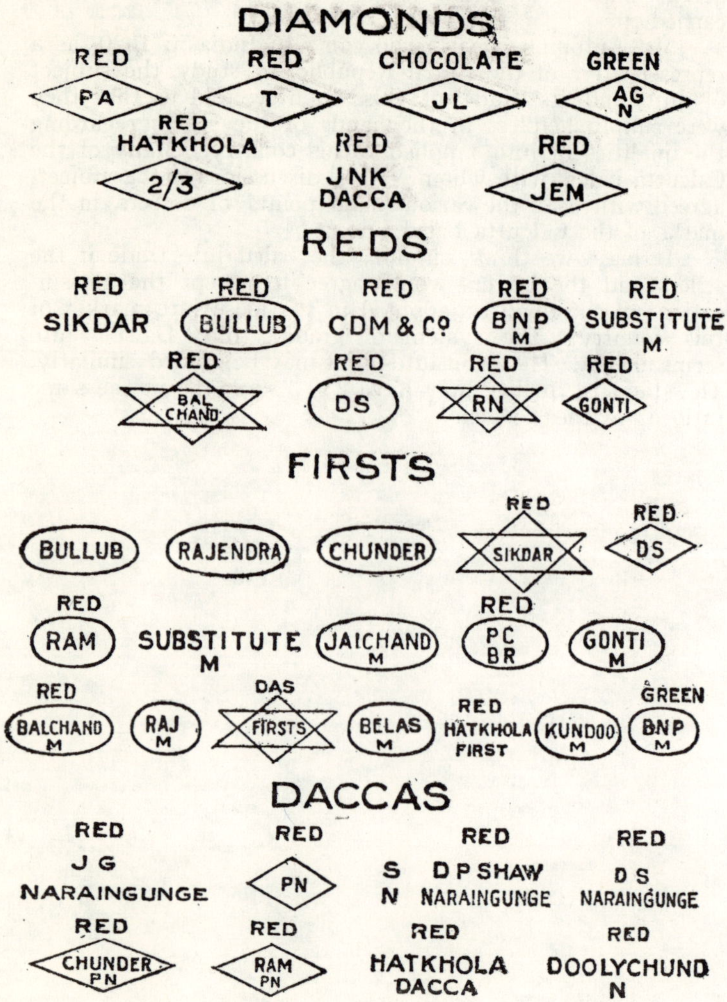
DIAMONDS
RED
PA
RED
T
CHOCOLATE
JL
GREEN
AG
N
RED
HATKHOLA
2/3
RED
JNK
DACCA
RED
JEM
REDS
RED
SIKDAR
RED
BULLUB
RED
CDM & Co
RED
BNP
M
RED
SUBSTITUTE
M
RED
BAL
CHAND
RED
DS
RED
RN
RED
GONTI
FIRSTS
BULLUB
RAJENDRA
CHUNDER
RED
SIKDAR
RED
DS
RED
RAM
SUBSTITUTE
M
JAICHAND
M
RED
PC
BR
GONTI
M
RED
BALCHAND
M
RAJ
M
DAS
FIRSTS
BELAS
M
RED
HATKHOLA
FIRST
KUNDOO
M
GREEN
BNP
M
DACCAS
RED
J G
NARAINGUNGE
RED
PN
RED
S
N
D P SHAW
NARAINGUNGE
RED
D S
NARAINGUNGE
RED
CHUNDER
PN
RED
RAM
PN
RED
HATKHOLA
DACCA
RED
DOOLYCHUND
N

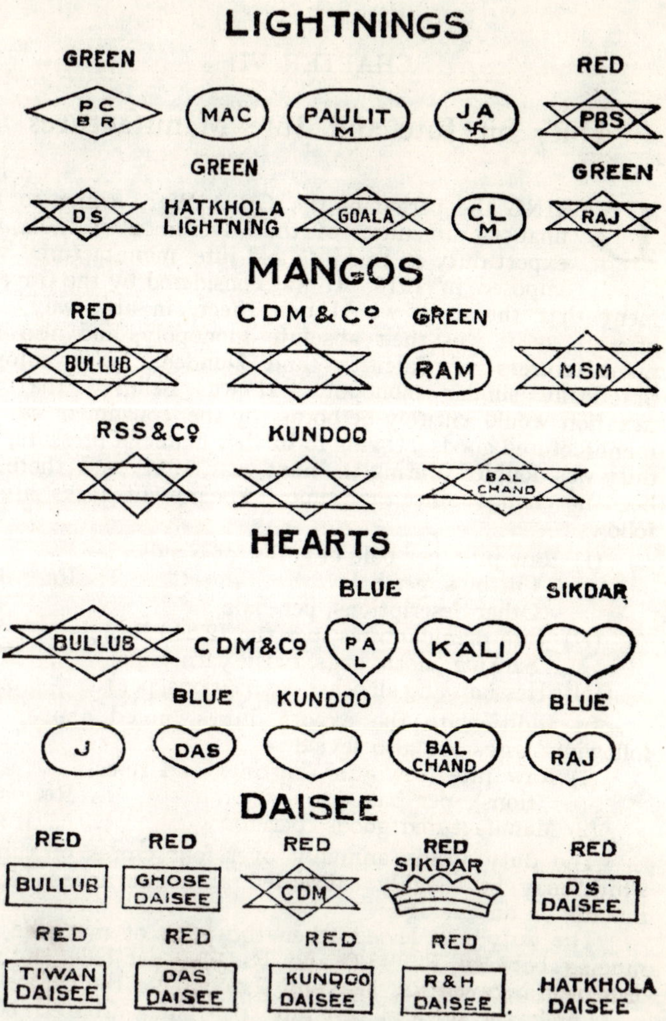
LIGHTNINGS
GREEN
PC BR
MAC
PAULIT M
JA
RED
PBS
GREEN
DS
HATKHOLA LIGHTNING
GOALA
CL M
GREEN
RAJ
MANGOS
RED
BULLUB
CDM&Co
GREEN
RAM
MSM
RSS&Co
KUNDOO
BAL CHAND
HEARTS
BULLUB
CDM&Co
BLUE
PA L
KALI
SIKDAR
BLUE
J
DAS
KUNDOO
BAL CHAND
BLUE
RAJ
DAISEE
RED
BULLUB
RED
GHOSE DAISEE
RED
CDM
RED
SIKDAR
RED
DS DAISEE
RED
TIWAN DAISEE
RED
DAS DAISEE
RED
KUNDOO DAISEE
RED
R&H DAISEE
HATKHOLA DAISEE

CHAPTER VII

Duty on Jute and Jute Manufactures

DURING the period of the Great War, owing to the financial difficulties of the Government of India, the export duty on raw jute and jute manufactures was imposed, in 1916. It was considered by the Government that the duty would not affect, in any way, the growers, as it was their absolute monopoly, and also the manufacturers of Calcutta and Dundee, who enjoyed practically similar monopoly. It was believed that the taxation would entirely be borne by the consumers of the manufactured goods. Owing to further financial pressure, the duty was doubled, during the next year, 1917, and there has been no change since that time. The present rates are as follows :—

(I) Raw jute, per bale (400 lbs. or 5 mds.) :—
Cuttings, per bale ... ... Rs. 1 - 4
Other descriptions, per bale ... ,, 4 - 8
(II) Jute manufactures, per ton (2,240 lbs.) :—
Sacking (cloth, bags, twist, yarn, rope) ,, 20 - 0
(III) Hessians and all other descriptions, per ton ,, 32 - 0

In addition to the export duties, noted above, the following cesses are also levied :—

(1) Raw jute (raw jute, cuttings and rejections), per bale ... ... Re. 0 - 2
(2) Manufactured goods, per ton ... ,, 0-12

The duty yields annually over four crores of rupees, which may be roughly calculated at about annas 14 per maund, on an average.

The duty was levied when the price of raw jute was ranging between Rs. 48-0 and Rs. 53-0 per bale, and that of hessian between Rs. 17-0 and Rs. 19-0 per 100 yards.

Lately the price of raw jute, per bale, dropped below Rs. 30-0 and that of hessian below Rs. 10-0. If the baled jute and the gunny markets, which are controlled by organised rich merchants, could not stop the course of the depression of particular goods in which they practically enjoy monopoly, there is no wonder, that the poor illiterate

and disorganised cultivators of the crop, who cannot hold off the sale even for a week, have had to suffer a severe loss ; as a matter of fact, they had to sell the crops of 1930 and 1931 at less than half the cost of production. Although the crop of 1931 was severely reduced, more or less by a half, the markets remained as dull as they were in the previous year of the excess production. The price rose, of course, to a very small extent, when all the crop passed off the hands of the growers. By the end of the month of October, 1931, more or less 75 per cent. of the crop was sold off by the raiyats.

The present trade depression prevails throughout the world. The Great War has greatly impoverished all the nations, directly or indirectly. It will take a long time to bring about normal conditions. Every country demands cheaper commodities. It is stated that some countries are seriously making attempts to introduce such cheap material as paper, rejected cotton or other fibres, as a substitute for jute, for sacking and wrapping purposes. In these circumstances, the export duty on jute and jute manufactures is not at all safe for the crop. It may also be considered that when the duty was imposed, the prices of raw jute and jute manufactures were high, and the duty was then, at about one anna per rupee, on raw jute and about three pice per rupee, on the hessian, but during the seasons of 1930 and 1931, it was about 3 annas per rupee, on the raw jute and an anna and-a-half per rupee, on the hessian. The rates, in the circumstances, appear to be very heavy, coming to about 20 per cent. on raw jute. If the duty be not abolished totally, on account of the financial troubles of the Government, it should be reduced proportionately.

It is to be regretted that up to the present, no amount, out of the revenue derived from the duty on exports of raw jute and jute manufactures, has been spent for the benefit of the growers or for the improvement of the crop. A considerable portion of this revenue should be utilised for the construction of roads, excavating and cleaning canals and tanks, and also opening *golas* for sale of the cultivators' crops, at the most favourable rates.

CHAPTER VIII

Famous Jute Markets

Narainganj

THE Narainganj market is supplied with jute, in the beginning of the season, from Bikrampur in the district of Dacca, and the adjoining *char* lands in Dacca and Tipperah which are annually inundated. This jute resembles mainly the District quality.

As the season advances, jute from the interior of Mymensingh and Tipperah begins to come in. The high-land jute of these districts is exceptionally good in quality, and its colour and gloss do not fade for a long time in storage. The fibre of this market is strong and free from fraudulent watering, which destroys its natural colour and gloss, even after it is thoroughly dried for storage. The peculiarity of this quality of Narainganj jute is that it is hard and strong, and bears more tension in spinning. The season closes here in February, when jute containing sacking-warp only is available. It is difficult to keep up the guaranteed warp of 60 per cent. during this period of the season, when a separate assortment is to be made.

The cuttings obtained from Narainganj jute are the longest, and best for use in the mills, as hessian-weft. Narainganj jute is mixed with other qualities, in the following proportions, in batching of fibres :—

Narainganj (strong)	2 parts
Serajganj	1 part

or

Narainganj (strong)	2 parts
Serajganj	1 part
Haldibari (soft)	1 ,,

or

Narainganj (strong)	2 parts
Madaripur (hard-strong)	1 part
Serajganj (soft)	1 ,,

Dacca

This quality of jute is obtained from parts of Dacca and Mymensingh districts. It comes into the market early in the season and is available for four months only, till the end of October. After October, the Dacca purchasers get their supply from Mymensingh district, both by river and rail.

The white colour of the fibre of low-lying tracts changes to reddishness, after three or four months' storage. It is a strong fibre.

From the latter part of the season, Dacca supplies the quality similar to that of Narainganj, *i.e.*, strong and sound jute.

Serajganj

In the beginning of the season (*i. e.*, July), local jute called *Deswal* is imported here. It is a medium fibre of both white and blackish colours, neither very fine nor thick. As the season advances, jute, from the other side of the river Jamuna—Panibari, Sarishabari, Porabari, etc.,—comes in. This quality of jute, which is different from the local lowland jute, is very good in colour, strength and fine texture, yielding 50 to 70 per cent. hessian.

Afterwards, jute, from Bogra and the Southern part of Rangpur, is imported. From October, *Deswal* and Mymensingh qualities are getting scarce, while the market is mainly supplied with the jute of parts of Northern Bengal, Assam and Cooch Behar, carried by boats.

The Northern jute is generally of very good quality, although without any particular colour. It is soft and bears more twist. The jute grown by the river Teesta (a tributary to the Brahmaputra) is strong and fine. A limited quantity of *C. olitorius*, called *Tosha* jute, is available at the Serajganj market.

For convenience of communications, jute is sold at Serajganj throughout the year. In importance, as a jute market, Serajganj occupies the second position, next to Narainganj. Fraudulent watering is practised, in the Serajganj local qualities.

Northern Markets

The jute, from Northern districts, comes into the markets, somewhat late in the season. Before September, the supply is limited. The jute of these tracts is fine and soft, consequently the Calcutta mills use this quality for batching purposes, in mixture with other hard and strong qualities. The North Bengal jute is soft and somewhat weak.

Purnea

Purnea jute is not included in the quality of the Northern jute. It is inferior in colour, strength and length. White stuff is rare. It is grown generally on low lands. Kissenganj, Barsoi, Kasba and Forbesganj are the principal markets in this district. A better class of jute is however imported into the district, from Morang, in Nepal State.

Madaripur

The Madaripur jute is mainly grown in the low country, washed by the Ganges. In most parts of these tracts, silts are deposited annually. It is a strong and coarse fibre of inferior colour. This quality, on the whole, is considered to be the best for sacking warp. The low grade jute of the Madaripur Division, with or without tap-roots, is called *Dowrah* jute.

Madaripur market is opened early in the season. The crops of the districts of Faridpur, Backerganj, and parts of Khulna are sold here.

The practice of fraudulent watering is usually more extensive, in this Division, than anywhere else. "Heart damaged" is the common complaint against this jute. Sometimes the "Heart damaged" jute of Madaripur, when opened, would give nothing but dust as its contents. For this reason, the greatest care is to be taken to examine the bales imported from Madaripur.

Of late, the quality of the Madaripur jute has greatly improved. The malpractice of fraudulent watering is coming to an end. Formerly, the raiyats used to uproot the plants for harvesting, instead of cutting at the roots. Cutting is now widely practised.

For the improvement of jute of this Division, the recent introduction of thinning the plants in the fields, some time before the flowering stage, is commendable.

Cuttack

Cuttack quality of jute comes from Orissa. It is a coarse (thick) but strong jute, suitable for sacking warp when free from sticks (broken stalks). The fibre of the Cuttack Division was, until recently, more or less sticky. This defect is being steadily removed.

CHAPTER IX

List of Jute Markets

A. Markets dealing in " Jat " Quality of Jute

Mymensingh
Kaoraid (Mymensingh)
Shambhuganj ,,
Jamalpur ,,
Sherpur ,,
Pearpur ,,
Ballipara ,,

Nicklidampara (Karimganj)
Mymensingh
Gouripur (Mymensingh)
Deonganj ,,
Dacca
Narainganj (Dacca)
Akhaura (Tipperah)

B. Markets dealing in " District " Quality

Elashin (Mymensingh)
Ballah ,,
Naga ,,
Panibari ,,
Sarishabari ,,
Nandanpur ,,
Shaturia ,,
Bahadurabad ,,
Bhairab Bazar ,,
Benanai ,,
Soya ,,
Bilasipara ,,
Babuapara ,,
Kedarpur ,,
Jamurki ,,
Ratanganj ,,
Mirjapur ,,
Porabari ,,

Sreenagar (Dacca)
Lohaganj ,,
Ghior ,,
Kanchanpur ,,
Kalliganj ,,
Baira ,,
Shinghhair ,,
Chilmari (Rangpur)
Chandpur (Tipperah)
Ashuganj ,,
Pabna
Dewantala (Bogra)
Fulchhari ,,
Pabna
Nagarbari (Pabna)
Nakalia ,,
Berah ,,

C. Markets dealing in "Northern" Quality

Haldibari	(Rangpur)	Shookanpukur	(Bogra)
Domar	,,	Mahimaganj	,,
Nilphamari	,,	Santahar	,,
Saidpur	,,	Chilhati	,,
Rangpur	,,	Jaypurhat	,,
Kurigram	,,	Dinajpur	
Durwani	,,	Raiganj	(Dinajpur)
Tulshighat	,,	Jaffarganj	
Naldanga	,,	Mirganj	
Alipur	,,	Jalpaiguri	
Rajshahi		Siliguri	(Jalpaiguri)
Raninagar	(Rajshahi)	Cooch Behar	
Atrai	,,	Mathabhanga	(Cooch Behar).
Khansamahat	(Bogra)		

D. Markets dealing in "Madaripur" Quality

Barhamganj	(Faridpur)		
adaripur	,,	Panchuria	(Faridpur)
Goalah	,,	Goalundo	,,
Rajore	,,	Rajbari	,,
Khejurtala	,,	Belgachi	,,
Patarhat	,,	Pangsha	,,
Idilpur	,,	Osmanpur	(Nadia)
Faridpur		Kumarkhali	,,
Khankhanapur	,,	Kushtia	,,

E. Markets dealing in "Non-Standard" Qualities

Kissenga r.	(Purnea)	Forbesganj	(Purnea)
Barsoi	,,	Azimganj	(Murshidabad)
Purnea		Jessore	
Kasba	,,	Cuttack	

F. Markets dealing in " Daisee " Quality

Baidyabati
Sheorafulli
Belgachia
Kristaganj
Chanditalla
Baduria
Chanduria
Magrahat
Kalna

The Calcutta Bazar includes the following markets where business of loose jute, chiefly of Madaripur Division, is carried on :—

Hatkhola
Ultadanga
Shambazar
Baghbazar
Chitpore
Fulbagan

CHAPTER X

Review of Prices

THE export of jute, to foreign countries, commenced in 1828. Before that time, it is said to have been sold at Re. 1 per maund, at its highest ; the ordinary rate was one pice a seer. Ever since that time the price of jute has been steadily rising. During 1845, the prices recorded, were at Rs. 9-4 in January, and at Rs. 8-10 in July, per bale (5 maunds), in Calcutta. The price of picked quality reached Rs. 30, per bale, in 1882. Then there was a fall-off to Rs. 25, during a period of five years. The market again steadily rose till 1894, when the picked quality was sold at Rs. 44 per bale. There was again a fall, which continued till 1904, when the picked quality was sold at Rs. 36 per bale. Since then, the market was again on the rise up to the year 1913. The maximum price, per bale of ordinary quality (*M*), reached Rs. 69. The market continued to be dull from 1914 to 1919, owing to restrictions on exports, during the period of the great European war, and to shortage of tonnage and other difficulties following, after declaration of peace. A much better condition of the jute market again prevailed from 1920 up to 1929. The highest price reached was Rs. 25-8, per maund of loose jute, or Rs. 114-0, per bale of *M* quality, during 1925. Excess production of the raw fibre and that of the mills, during the year 1930, and general trade depression, were the causes of an unprecedented fall-off in the prices of the raw fibre and the mills' products. The price of baled jute dropped from Rs. 66 to Rs. 30, loose jute from Rs. 12 to Rs. 3-8, and hessian from Rs. 18 to Rs. 9. During the year 1931, the field production was reduced to one-half of the previous year's crop and the mills' production was also considerably shortened, still the markets continued to be very dull. Nobody should expect, owing to the world-wide trade depression, and the introduction of cheaper paper baggings in America, that the markets would revert to such a favourable state as that prevailed before.

The period, from July to December, is the busy season of the jute trade, and the first six months of a year, from January to June, are the dull season. Often, however, the market is stronger, during the dull season, than in the busy season, when the stock in hand runs short, and the demand for the fibre is still keen. New crop begins to come into the markets in July.

CHAPTER XI

Railway and Steamer Tariffs

CALCUTTA is the centre of jute trade in India. Jute is carried from districts to Calcutta by Railways (about 50%) and Steamers (about 47%), the remaining portion being conveyed by country boats (about 2%) and bullock carts. Of the Railway traffic in jute, E. B. R. is responsible for carrying about 90%, A. B. R. 8%, and E. I. R. and B. N. R. combined 2% only of the total goods.

In India, goods are principally divided into 10 classes, and freight is charged, according to a class rate, as follows :—

	Rate, per maund, per mile.	
Class.	Minimum, pies.	Maximum, pies.
I	0·100	0·38
II	0·100	0·42
III	0·166	0·58
IV	0·166	0·62
V	0·166	0·77
VI	0·166	0·83
VII	0·166	0·96
VIII	0·166	1·04
IX	0·166	1·25
X	0·166	1·87

Machine pressed jute and hemp are classified, as 2nd class goods, and those, which are not machine pressed, are charged as 4th class; while jute seed and gunnies are charged, at the 1st class rate.

Referring to the tariff books of different Railways, it appears, that machine pressed jute and hemp are carried, over B. N. R. and E. I. R., at the 2nd class minimum rate of 0·100 pie per maund, per mile, in wagon loads, not less than 130 maunds in a consignment, for a distance not below 120 miles. But there is no jute on these lines, beyond 50 miles from Calcutta. Unfortunately, E. B. R. does not accept jute to carry at the minimum rate of the schedule. There

are special rates, between different stations (*vide* Statement on E. B. R. and Steamer Tariff). Machine pressed jute is usually charged, at the 2nd class maximum rate of 0·42, per mile, per maund, even for the longest distances, with the exception of that despatched from Mymensingh to Calcutta, which is charged at 0·389 pie. It is also not understood as to why a consignment of jute is carried over E. B. R. and Steamer Services, from Dacca to Calcutta, *via* Narayanganj and Khulna, extending over a distance of 322 miles, instead of *via* Narayanganj and Goalundo which shortens the distance by 52 miles, and should reduce freight considerably.

There is much to criticise in the way in which the authorities of the Steamer Companies fix freights from one station to another. To them, distances appear to be of no consequence. For example, they charge the same freight for jute booked direct for Calcutta (Nimtalah Ghat), either from Serajganj (544 miles), or from Madaripur (415 miles), or from Chandpur (455 miles), or from Narayanganj (492 miles). The steamer freight for jute from Narayanganj to Calcutta, by direct Steamer Service, is Re. 1-14-10 for each bale of 3½ maunds, while that from Dacca to Calcutta, by combined Railway and Steamer Services, is Re. 1-12-3, for the same quantity (*vide* Statement regarding E. B. R. and Steamer Tariffs).

The freight is charged, by the Steamer Companies, at about 8 annas per maund (Rs. 12 or Rs. 13 per ton) of machine pressed jute, for carriage between Madaripur and Calcutta, while country boats carry the same at 3 annas per maund of machine pressed jute, and at 4 annas per maund of loose jute, in drums. The charges by country boats, from Narayanganj to Calcutta, would not be more than 1 anna higher than that charged from Madaripur to Calcutta. Country boats take 7 to 10 days to complete the journey, during the rainy season. But still the merchants prefer steamer service to the country boats, for fear of accidents. That should however be no reason, for the companies, to fix inequitably high rates, on the most important article of commerce, in Bengal.

The charges of the Indian Steamer Companies,* on coast line traffic, are also very high, in comparison with the rates charged by Foreign Steamship Companies (*vide* Shipping Freight Schedule). The freight, for a ton of baled jute, from

* The British companies who have been granted charter to do trade in India, are included as Indian companies. These British Companies carry on practically all inland and coastal traffic.

Calcutta to Hamburg or to Antwerp, in Europe, is 27*s*. 6*d*. (equivalent to about Rs. 14) and that from Calcutta to New York or Boston or Philadelphia, in America, is 32*s*. 6*d*. (about Rs. 17), while the Indian Steamer Companies charge similar freightage, even for manufactured goods, from Calcutta to Rangoon or Colombo, Madras, Bombay, etc. It may be noted that special reduced rates are offered for a full shipload of consignment by the Shipping Companies of other countries. The Indian Railway and Steamer Companies should, therefore, revise and reduce the present rates.

Freight, both on raw material and its manufactured goods, is a factor for development of an industry of a country. The present distressed condition of the jute industry deserves specially favourable consideration, in regard to reduction of freight as proposed, both in and out of the country.

CHAPTER XII

Jute Dealers

THE business of selling and buying is effected by some classes of middlemen. The first of them are those who are called *beparies* or *farias*. They visit the cultivators, at their houses, settle the price, make weighment, and carry the fibre to the *mahajan*, a trader who advances money to the *beparies*, for purchase of jute. The *bepari* gets a commission of about 2 annas per maund. Some *beparies* have their own small capital. They are, for the most part, a class of illiterate men, who care little for honesty. By this class of dealers, the cultivators are cheated in weighment, the fibre is fraudulently watered and sold to the *mahajans*, or the local agents of the European firms of Calcutta. The *beparies* receive, very often, advances from the buyers, when they get the usual commission of annas 2 per maund. It has been the custom of the markets, in most districts, for the *beparies* to offer 1 to 2 seers of jute, extra for each maund, sold to the buyers as *dhaltha* (shortage in drying). But with all their faults, the services of the *beparies* are indispensable.

The *mahajan* is an Indian trader of substantial means. He must have a godown of his own, where he stocks the loose jute purchased. He makes the loose jute into drums or *katcha* bales and sends the jute either to an *aratdar* in Calcutta for disposal, or sells it to a local agent of a baler or a shipper or a mill. An *aratdar* is an important personage of the Indian trade. He sells goods sent to him by *mahajans* or *beparies*, on some occasions, at his own discretion, if no agents of their own are appointed by the sellers, to look after their sales. He gets a commission from the sellers. He is to make, from time to time, advances to his constituents, on terms settled between the parties.

Now appears a broker. Without his assistance no business can be effected satisfactorily. He must thoroughly be conversant with the market. He should have a perfect knowledge of different qualities. He must also be a person

of confidence. Both the sellers and the buyers rely on his honesty. The sale is entirely operated by his co-operation, between the sellers and the buyers.

Balers are the buyers of raw jute, who bale the jute by means of the press worked with steam power, for export.

The merchants, who export jute to other countries, are called the shippers. They purchase baled jute, as well as raw jute to be baled by them, for export. Many shippers have their own presses. If they have no presses of their own, they get the raw jute baled in the presses belonging to others, on the hire or contract system.

There are some organised associations of the Calcutta jute dealers. Each association looks after the interests of its members, in the trade. The members of an association are to abide by the rules and by-laws of the association. The following are the well-known associations in Calcutta :—

Indian Jute Mills Association

This Association was established in 1884. The aim and intention of the Association is to encourage united feeling and action to protect the interests of the mills.

All persons owning and managing jute mills or holding power-of-attorney to represent them in India, are eligible to join the Association, on payment of the annual subscription in advance, at the rate to be fixed at each Annual General Meeting. The subscription is calculated on the number of looms or spindles owned or managed by each member. The subscription was raised in 1931, from annas 6 to annas 14, per loom, per annum.

The Association was managed by a Committee consisting of a Chairman and four members, appointed at the Annual General Meeting, each year. During the year 1930, the number of the members of the Committee was increased from 5 to 7 including the Chairman. For the first time an Indian member, a representative of Messrs. Sir Sarupchand Hukumchand & Co., was taken on the Committee. The Secretary of the Bengal Chamber of Commerce is *ex-officio* Secretary of the Association.

The annual income of the Association, during 1929-30, amounted to Rs. 41,857-13-0, from subscription (Rs. 38,083-1-0) and interest (Rs. 3,774-12-0).

All the money received on account of contribution, subscriptions, fines, etc., is applied towards carrying out the following noted objects :—

(*a*) To encourage and finance technical developments in the plants and machineries, for the manufacture of jute and jute products.

(*b*) To encourage and finance scientific exploration of new uses to which jute can be applied and of discovering bye-products.

(*c*) To open out new markets.

All the Calcutta Mills, except the seven following, have joined the Association :—

Adamjee Mills.
Gagalbhai Mills.
Sree Hanuman Mills.
Premchand Mills.
Agarpara Mills.
Ludlow Mills of Messrs. Ludlow Jute Co., Ltd.

Calcutta Baled Jute Association

The object of this Association is to regulate the Baled jute trade in Calcutta.

All firms or persons, interested or desirous to engage in the baled jute trade, are eligible as candidates for membership of the Association. The candidates, for membership, shall be proposed by one member of the Association and seconded by another. Elections shall then be made finally by the Committee, who may elect or decline to elect a candidate, as a member. The members are to abide by all the rules and by-laws of the Association then, and as from time to time, in force.

Each member shall pay an entrance fee of Rs. 500 on joining the Association, and the subscription of Rs. 32 annually, payable in advance in the month of September.

The members of the Association shall register with the Association a list and description of the marks and counter-marks of the baled jute they may offer for sale in the Calcutta market.

The members of the Association shall not do any business, in Calcutta or elsewhere in India, in any baled jute, cuttings or rejections, with or for any firm or person who is not a member of this Association, save and except with the Calcutta jute mills.

The general management of the business and affairs of the Association shall be entrusted to a Committee consisting of nine members, to include an equal number,

each, of the representatives of the Shippers, Balers and Brokers. The members of the Committee shall be elected by a majority of the votes of members, at the Annual General Meeting. The Committee shall elect their own Chairman. The Secretary of the Bengal Chamber of Commerce shall be *ex-officio* Secretary of the Association.

Jute Dealers' Association

This Association has lately been established, in Calcutta, by the European dealers in loose jute. The rules and regulations for membership and management of affairs of this Association are not known to us. The Secretary of the Bengal Chamber of Commerce is *ex-officio* Secretary of this Association.

Calcutta Jute Balers' Association

This Association was founded, in 1909, to provide rooms and other facilities to Indian Jute Balers, Shippers and Traders, for holding and conducting sales of property, brokerage or commercial transactions, etc., to promote, protect and watch over the general commercial interests, to collect and circulate statistics and other information relating to the trade, to adjust controversies between members, to arbitrate in the settlement of disputes, arising out of commercial transactions, and to maintain uniformity in rules, regulations and usage of trade. The Association was incorporated on the 8th June, 1918, under Section 26 of the Indian Companies Act of 1913. The affairs of the Association are managed by a Committee of four members, elected annually, by the members, at the Annual General Meeting, and a Chairman elected by the Committee. The routine office work is conducted by the Secretary.

Merchants and firms carrying on or desirous to carry on trade, in baled jute, in Calcutta, may be elected as members of the Association. A candidate for election, as a member, shall sign the prescribed application form and be proposed by one and seconded by another member, in the form itself, and shall deposit therewith a sum of Rs. 300, as admission fee, and Rs. 216 as the annual subscription in advance. The Committee may elect or decline to elect a candidate as a member of the Association.

East India Jute Association, Ld.

(Futures Market or Fatka Bazar)

The Association was incorporated on 3rd June, 1927, under the Indian Companies Act of 1913.

The Association intends to provide the jute trade with a well regulated and properly controlled futures market

working on scientific lines, to enable people engaged in the jute trade to enter into *bonâ fide* hedging transactions, against their commitments. In this system, a trader of small means is able to buy or sell goods worth many times above his means, owing to the credit facilities available, and make good profit, if he would carefully study the market.

The Association is supposed to protect merchants, against the risk involved, in the fluctuation of prices.

The commercial world has produced institutions known as " exchanges " where traders can buy or sell goods months ahead of the actual specified time when delivery is taken or given. Such institutions provide a continuous market both for the consumer as well as the producer. The jute market would have been more fluctuating, in the absence of a continuous market like one the Association represents, and a buyer would have sustained greater loss on a transaction. Although the futures markets are established, on a speculation basis, they hold a practical influence over the markets to minimise scares and to reduce risks of *bonâ fide* traders. It cannot be denied that there are evils of speculation in such markets, which may however be checked by admitting persons of sound financial standing as members, in order to enforce the right of delivery on buyers or sellers, according to contracts. According to the rules of the Association, 250 bales of jute of specific marks are the minimum to be dealt with in a sale. In a dispute regarding quality, the buyer has the right to claim arbitration on the purchased goods.

If the Association be thoroughly organised with an efficient staff for getting into effective touch with the cultivators and the crop, also with the demand of the world, and if it help the buyers and the sellers with information relating to futures market by issuing now and then leaflets to the members, for checking ruinous speculations, it would really be highly beneficial to the members and the trade.

The members of the Association pay a subscription of Rs. 140 annually.

The Association was founded by Mr. H. P. Bagaria, a Marwari jute merchant of Calcutta, with a fund of Rs. 45,000 raised by subscriptions by the members. At the end of the third year, on 30th June, 1930, the fund accumulated Rs. 1,19,208, showing a rapid and steady progress of the institution.

Mr. Bagaria, a man of great ability and business insight, is to be congratulated on the success of the Association.

PART V—INDUSTRY

CHAPTER I

History of Modern Jute Industry

THE history of modern jute industry in Bengal is most interesting. Although jute had been grown in Bengal for centuries and woven into such coarse fabrics as bags and mattresses, the industry was not introduced into Europe, on a commercial scale, prior to 1833. It was at Dundee, which had for a considerable time been one of the most important seats of linen manufacture, that the jute industry was installed.

Thomas Neigh, a Dundee merchant, received in 1820 a small consignment of jute from a London firm. He tried to induce some of the flax spinners of Dundee to try the fibre over their machinery, but they were not convinced that it would be successful, until at length, he got Messrs. Bell and Balfour, flax spinners, to consent to make an experiment with the fibre. The experiment, which was conducted sometime between 1825 and 1826, was not successful, so as to spin it into yarn, and some years elapsed before any further attempts were made. Mr. Neigh was a man of great perseverance. He did not lose heart. In 1832, he obtained another consignment from London and succeeded in persuading the same spinners, Messrs. Balfour and Meldrum, successors to Messrs. Bell and Balfour, to try the experiment over again. This time they were successful to a great extent. From about the end of that year, small quantities of jute were regularly spun by them. Other parties followed gradually and the industry was firmly established there, after the machineries were suitably adjusted. By the year 1850, the use of jute was extensive at Dundee. Dundee still maintains its leading position in the world's jute industry.

In Bengal, the manufacture of jute, by application of European machinery, was not introduced before 1854, when Mr. George Ackland, a coffee planter of Ceylon, founded the first mill at Ishra, near Serampur, in the District of Hoogly, which is now known as the "Wellington

Mills." The second jute mill was established, in 1857, by the Borneo Company, Limited, now known as the "Baranagar Jute Mills." In 1863-64 the "Gouripur Jute Factory" came into existence. Since then, in rapid succession, jute mills have sprung up in and about Calcutta, on the river Hoogly. Now they number not less than eighty-nine.

There were 43 mills at work in Bengal and one in Bombay, during 1907. The number of looms working in 1887 was 7,164, which increased to 9,841 in 1895, to 23,884 in 1905, and to 39,401 in 1917. In 1920, 76 mills were in working order and 16 new mills under construction. Out of the 76 running mills, there were three in Madras, one in the United Provinces of Agra and Oudh, and the rest (72) in Calcutta. The capital laid out for these 76 mills up to 1918-19 amounted to Rs. 14,07,14,680 registered in India and to £2,209,716 registered in the United Kingdom, including Rs. 3,42,90,500 and £71,358 respectively in debentures. During 1919-20, fourteen mills were registered in India, with an authorised capital of Rs. 6,62,00,000, showing a great development of the industry in Bengal. The actual consumption of jute by these mills nearly doubled during the period between 1910 and 1920 ; while the foreign exports of raw jute continued to increase very steadily, each year exceeding the preceding one, up to August 1914, when the great European War broke out. War conditions, however, restricted the exports of raw jute, even after peace had been declared in November 1918, and the Government restrictions were withdrawn, owing to shortage of mercantile tonnage and other after-effects of the war. There was again an all-round expansion of the trade, from 1922, which continued till 1929. The number of mills increased, in Bengal, from 72 to 89. There has again been a very serious set-back, since 1930, both for the raw product and the manufactured goods, due to the excess of production of raw fibre as well as that of the mills, and also to the unprecedented world-wide trade depression.

The year 1930 was not only distressing to the cultivation alone, but to all branches of the industryt The production of raw fibre was appreciably extending from 1926-27, which produced 123·8 lakh bales, agains. 93·5 lakh bales during the preceding year. From 1927-28 to 1929-30, the outturns were 111·5, 104·8 and 110·0 lakh bales respectively, although not apparently affecting the trade very much before 1929-30. In the meantime, the world's buying capacity was steadily depreciating. The

Indian mills, on the other hand, introduced, since July, 1929, 60 hours working a week, instead of the usual 54 hours work, and this system was allowed to continue for nearly a year, thereby producing much more than was required. Heavy stocks accumulated in the mills' godowns. The result was that the price of raw jute fell to Rs. 3-8, which was about half the cost of production, although the production of raw jute in 1931 was reduced by one half. The price of *pacca* bale (400 lbs.) of *firsts* fell from Rs. 66 to Rs. 30 or less towards the close of 1932. Similarly the price of the manufactured goods dropped down by about 50%. Since October, 1930, the mills' production has been considerably curtailed by working 40 hours a week.

We are hardly competent to forecast how long this depression may continue. It is understood that post-war exceptionally high prices of raw and manufactured goods forced America to replace jute by some other fibres, and, as a result, paper bags are being introduced in large numbers, specially for cement packing. If paper bagging be a success, then the jute industry may have a severe shock (*vide* the Chapter on substitutes).

Unlike the Indian capitalists of Bombay who own nearly all the cotton mills of that place and staff them with Indians, Bengalees unfortunately possess one mill only, initiated by Raja Sreenath Roy and Brothers; they are also not found amongst the supervising staffs of the mills. It is however satisfactory that among the enterprising Marwari merchants of Calcutta, Sir Sarup Chand, Messrs. Birla Brothers and Messrs. Soorajmull Nagarmull & Co. floated three jute mills. There are two other mills lately established by Indians, Messrs. Adamji Hazi Dawood & Co. and Messrs. Mafatlal Gagalbhai & Co. Before these belated undertakings, the Maharaja of Darbhanga was the only Indian owner of a small jute mill at Balliaghata, in the suburb of Calcutta. Except some three mills possessed by Americans, known as the Ludlow's, all other mills were floated with British Capital.

Jute is at present used more than all other vegetable fibres combined, except cotton. It is the cheapest and most easily spun of any of the soft fibres and it is most adapted for purposes where strength and durability are of secondary importance. Jute has replaced hemp for many uses, such as covering for cotton and wool packing, packages in transit, sacks for coffee, sugar, grain, etc., and its popularity is due to its cheapness.

The value of exports of raw jute and jute manufactures, from India, was most satisfactory for a number of years, from 1926-27 to 1929-30, amounting to about 83 crores of rupees per year, on the average; while that of cotton and cotton goods averaged, per year of the same period, about 68 crores of rupees.

Unfortunately for the jute industry, the value of raw jute and manufactures fell suddenly to 44·74 crores of rupees, during 1930, and the prospect of recovery appears to be gloomy. In spite of this extremely abnormal fall-off in the value of the exports of jute and jute products, it still ranks second to cotton, in importance, in the export trade of India.

CHAPTER II

Indian Manufactures

THE manufacture of jute fabrics by hand is an old industry of Bengal. Sir George Watt, a great authority on economic products of India, says that jute cloth was used for clothing by the poorer classes a century ago. Since then the cheap and good-looking European cotton goods have completely replaced its use in this country.

The principal articles of hand-loom manufacture of India were gunny bags, mattings, rugs, and sails for country boats. It is said that paper was also manufactured out of jute, in several districts, in North and East Bengal. The Indian weavers used to dye the fabrics in red, black and yellow colours of their own.

Prior to 1857, the export of jute manufactures from India represented hand-loom fabrics. In 1850, these manufactures were valued at Rs. 21,59,780 (£215,978). The high price obtained from the export of gunny bags considerably increased the resources of the Bengalee weavers who were chiefly cultivators. In the meantime, however, an enterprizing Scotch merchant introduced the jute-industry in the hemp manufacturing factories at Dundee, worked by steam power, which began to compete with manual labour in India, and in due time, since 1854, the hand-loom industry of jute rapidly declined. During 1880-81 the value of the total exports of manufactured jute amounted to Rs. 1,13,06,716, of which hand looms accounted for Rs. 2,69,553 ; and in 1887-88 the hand-loom manufactured articles exported were valued at Rs. 89,220 only, out of the total exports to the value of Rs. 1,15,18,577. The export of hand-loom manufactured articles has since entirely ceased. Thus Dundee assumed the leading position of a manufacturing town, while Bengal remained as a country producing raw fibre only.

The old hand-loom industry of jute in India is now well-nigh defunct. Gunny bags are woven in some districts to a very limited extent for the weaver's own use

or for his neighbours. The only industry still left in the hands of the people of this country is the manufacture of ropes required for everyday home use. A Bengalee family uses, at least, half a maund of jute a year for domestic purposes.

In the factories set up in India, jute is chiefly manufactured into hessian cloth and gunnies. *Rejections* and *cuttings* of jute were formerly used by paper mills alone; they are now utilised also by the jute mills in the manufacture of sackings. Coarse and strong qualities of the fibre are used in some steam rope works for cordage.

During the year 1917, there were 76 jute mills in India, out of which 72 mills were set up in Bengal with 39,401 looms—23,426 being hessian looms and 15,975 sacking looms, and in 1929 there were 89 mills with 51,440 looms and 1,084,130 spindles working, for which 329,291 persons were employed on a daily average. The total quantity of coal consumed by the mills during 1930 amounted to 827,000 tons. The mills are situated within a radius of about 50 miles in the neighbourhood of Calcutta. The Jute district is extending northwards from Calcutta on both banks of the river Hoogly.

On an average, there are about 500 looms and 10,500 spindles in each mill employing about 3,000 labourers. Each loom requires roughly 6 hands for spinning and weaving, but an attempt is being made to reduce the number to 5, with progress in efficiency. Each hessian loom (about 62% of total looms) consumes approximately 1½ maunds of jute per diem, and each sacking loom (about 38%) 3½ maunds. Working 60 hours a week, each of the looms of the Calcutta mills consumed actually 532·5 maunds of jute during the year 1930. On a rough calculation, 30 maunds of jute produce 1 ton (1 ton=27·22 maunds or 27 maunds and 9 seers; 1 maund =82 lbs. and 2 ozs.) of finished goods. Roughly speaking, 21 spindles supply yarn for each loom During 1918-19 the value of the jute manufactures exported amounted to about £35,101,466 against £8,480,052 for the exports of raw jute, which rose to £51,926,800 and £27,173,800 during 1929, respectively, but suddenly dropped to £31,858,000 and £12,884,700 during the following year, 1930.

From the statement tabulated below, it will be seen to what magnitude India suffered, from the effects of war, in its income from the exports of raw jute. This naturally considerably affected the interests of the cultivators of the country.

Values of exports of raw jute and manufactures

Year.	Value of raw jute in sterling.	Value of manufactures in sterling.
1913-14	£ 20,550,929	£ 18,848,759
1914-15	,, 8,606,802	,, 17,213,440
1915-16	,, 10,428,024	,, 25,318,934
1916-17	,, 10,858,736	,, 27,781,156
1917-18	,, 4,302,559	,, 28,562,050
1918-19	,, 8,480,052	,, 35,101,466

An immense quantity of gunny bags and jute cloth was sold by the Indian mills during the war. The uses of gunny bags for war purposes proved to be invaluable. They were utilised not only as sacks for carrying rations or other goods, but largely as sand-bags, for the defensive operations resorted to by the belligerent allies. There had also been a great expansion in the trade of jute fabrics in this country, the value of which was estimated, on an average, at Rs. 3,73,14,100 per year, during this period.

CHAPTER III

European Manufactures

THE Jute weaving industry was firmly established at Dundee in 1838. For a long time Dundee was the only manufacturing town in Europe. Later, the industry sprang up in many towns in Europe and America, of which Hamburg, Dunkirk, Genoa, Trieste, New York and Philadelphia are the most important. Japan too has turned its attention to this industry. Up to 1918, the manufactures of jute products had been limited to one factory in Japan at a place near Kobe, and another in Formosa. The German jute industry dates from 1861, the prejudice existing against jute at the time being so great that it was found difficult to raise the necessary capital. At about the same time, the mill industry of jute was also established at New York. Brazil did not lag behind to introduce it into her existing mills in Sao Paulo, where indigenous fibre paco-paco was used for the manufacture of bags for exporting coffee, etc. After the Great War, jute has been more extensively used in some of the Baltic States, substituting or mixing with hemp fibre in the manufacture of carpets as a home industry.

The position of the industry in European countries in 1929-30, where jute is either spun or manufactured to a considerable extent, is that there are about 100 mills in Scotland, England has 12, Italy 40, Germany 28, France 23, Belgium 18, Austria 12, Ireland 3, United States 60, Canada 10, Brazil 10, and Japan 6. Up to the present, Dundee still maintains its leading position in Europe.

Although the industry has been passing through a crisis never known before, India's monopoly in the raw produce of jute and cheap labour, still suggests that further development in the manufacture of jute will, before long, take place. It is India's interest to rise to the occasion to remove any defect in the fibre, and see that no deterioration should set in. Fraudulent adulteration of jute must cease, otherwise the industry is likely to suffer in the future. And it may be as well to note that persistent efforts are being made in

other countries either to introduce the cultivation of jute or to find a suitable substitute.

In European countries and America, jute is manufactured into shirtings, curtains, carpets and rugs. The great bulk of this fibre, however, is made into such commoner classes of cloths as hessian, tarpauling and sacking. Finer qualities are used for mixing with silk or for manufacturing imitation silk fabrics. It is also mixed with wool. Rejections and cuttings of jute are made use of in the manufacture of paper, and coarse but strong stuffs in cordage. The attempts at spinning higher counts than twenty have not been successful.

It has been applied extensively as a substitute for hemp ; for this purpose, the fibre is rendered soft and flexible by being sprinkled with water and oil. The jute is then left for 24 to 48 hours. After being squeezed by rollers and dressed with a hackle, the fibre becomes so beautifully soft and minutely isolated that it is suited for a number of purposes unknown a few years ago.

In the year 1820, jute was first experimented with in Europe, but the result was unfavourable, and, in consequence, brokers were required to certify that sales of hemp and other fibres, exported out of India, were not adulterated with jute. In 1832, in Dundee, experiments were again conducted with the fibre. The result was encouraging, and it was suggested that it might be used as a substitute for hemp. From that date jute gained rapidly in public favour.

Jute is one of the fibres that is capable of the most minute separation or subdivision, but only within the past few years has it been extensively used in the finer textile industries. For a long time, the difficulty of bleaching seemed insurmountable, and the trouble experienced in dyeing the material appeared likely to nullify every effort to utilise it. All these stumbling blocks have, at last, been removed. But its perishable nature, however, is fatal to its obtaining a position much higher than it has already attained.

The results of the chemical and microscopic investigation of jute, instituted by Messrs. Cross, Beaven and King, a few years ago, were that one sample experimented with was made to resemble *tasar* silk so closely that some care was necessary in distinguishing these substances, and another sample looked remarkably like wool. Before long, it is hoped that it should be quite possible to utilise the jute fibre for more useful purposes, by chemically altering its properties.

CHAPTER IV

Spinning

THE manufacture of jute consists of spinning and weaving ; but the raw fibre has to pass through various operations preparatory to spinning and weaving which commence with opening of the fibre of the bales, subsequent processes being batching, hackling, carding, etc., to render the fibre thoroughly split up and combed for spinning. It is not intended by us to deal with technical details in this work, but to give the trades people the information about the operations producing the goods in which they do business. These chapters have been written by the author from his personal observations at some mills.

Opening

A bale of the fibre is carried over to a machine called *Opener*, which removes the covering and the strings of the bale and opens the *heads* of the fibre. The fibre is then passed through the machine to make it soft and pliable, so as to be easily handled by the batchers. There is another machine called *Root-opener* which has a rubbing action. It opens and cleans the roots and removes dirt and sand through the grating underneath the cylinders.

The *kutcha* bales of jute of the market which are not hard pressed, as *pucca* bales intended for export, are carried to the Batching machine and opened by hand. In the Indian mills the opening machine is not required. But a machine like the *Root-opener* would be very useful for removing the dirt and sand often contained in the fibre, before it is lubricated in the Batching plant.

Batching

In the operation of *batching*, a mixture of oil and water is applied through a sprayer, over the fibre, when it is passed through the machine. The pieces of fibre are so placed on feed-arpons towards the holders of the machine that the top-ends of the fibre may first enter into the machine. This

operation lubricates and softens the fibre and considerably lessens the harshness and brittleness of it. Batching also makes the fibre pass easily through the subsequent processes of hackling, carding, drawing, etc., for spinning without lapping upon the rollers.

The batching liquid contains about 85% water and 15% oil and soap mixed in the following proportions :—

(*a*)	Whale oil	$2\frac{1}{2}$	gallons.
	Mineral „	2	„
	Seal „	$2\frac{1}{2}$	„
	Water	40	„
(*b*)	Seal oil	1	gallon.
	Sperm (whale) oil	5	gallons.
	Soap (washing)	10	lbs.
	Water	40	gallons.
(*c*)	Vegetable oil ...	6 or 7	gallons.
	(from Linseed, Cotton Seed or Mahua Seed, mixed together or separate)		
	Water	40	„

For common sacking wefts, a cheap batching mixture containing 10% of any lubricating cheap oil and 90% of water is recommended. In certain parts of India very cheap fish oil is available.

When jute is intended for bleaching, alkali in the form of soap, caustic soda or dilute ammonium hydrate is used to saponify the batching mixture. Soap is used in proportions of 3% to 5% of the oil and 1% of caustic soda.

Batching liquid is best prepared by measuring the oil into a tank and water in proportions of 3 : 1 being added. The mixture is then boiled, when the other portion of water is added ; and the whole again is brought to boil and kept heated until required. This liquor is applied to the fibre, at the rate of about 10 gallons to each 400 lbs. (one bale) of jute. In hot weather, a larger quantity of the liquid is wanted for the same quantity of the fibre.

Hand Picking

After batching, the sticks of the fibre are hand-picked, separating warf fibre from that of the weft and separately stacked for a day or two.

Snipping

Now the barky and mossy root-ends about a foot long from the bottom are cut off by labourers over a slightly curved sharp knife, 40″ long by 4″ in width, which is strongly adjusted over a wooden frame, about 2 feet high, placing the knife parallel to the floor. The finest and best quality of yarn is produced from the fibre of superior qualities, in which both the root and top ends are snipped off ; the root-ends being coarse and harsh and the top-ends being weak.

Hackling

After cutting, the fibre is carried to the Hackling machine, the root-ends being fed first. In the beginning of the operation, the sticks of the fibre are pieced and spread on the feed-arpons and then passed through over the pins of the machine which split up the fibre into finer filaments, producing rough slivers.

Hand-hackling is the process of sub-dividing or splitting up of the fibre of jute or hemp by pulling it in handfuls through the hackling tools instead of subjecting the fibre to the action of the *Hackling machine.* For pulling the fibre through the tools, the piece should be well spread out and drawn over the hackler by both the hands of an operator. By drawing the piece of the fibre by the hands and giving a number of blows to either side against the pins, it makes both faces of the piece equally dressed or combed. After the piece is thoroughly hackled, a lap is put upon it which is built with others into a compact bunch tied up with hands and stored until required.

A *Hand-hackle* consists of a piece of heavy wood, commonly $16\frac{1}{4}''$ long, 15″ broad and $1\frac{5}{8}''$ thick, of which $9\frac{1}{4}''$ by 8″ in the centre of the stock is studded with steel pins 6″ to $7\frac{1}{4}''$ long tightly driven into the stock. There are generally 11 pins in a row lengthwise and 5 parallel rows in all.

The *Hackling* operation is followed by processes of *Breaking, Carding, Drawing* and *Roving* preparatory to *Spinning.* All these processes are similar, with different arrangements, preparing slivers more and more uniform in size, quality and weight, and finally for spinning.

Breaking

The fibre is further broken in this machine and turned into rough slivers.

Carding

The primary function of *Carding* is the laying of the rough sort of slivers received from the Breaking machine parallel to each other and passing through the *Carding machine* to produce more uniform slivers.

Drawing Frame

Now the slivers have to be drawn through the *Drawing Frame I*, and also again through the *Drawing Frame II*. In these processes, the slivers are further split up and turned into finer filaments of more equal sized slivers, before they are taken over to the Roving machine.

Roving Frames

Over the *Roving Frames*, the fibre is once more split up where slivers are finally prepared. It is slightly twisted here. In doing so they take a sliver from the Drawing frames and give it a further combing.

The Spinner

The partially twisted thread is at last spun into yarn, which should be of uniform size and of maximum strength throughout, for a particular type and for the purpose of particular use. The yarn that is produced by 20 or 21 spindles feeds one loom in the operation of weaving. Two sets of 80 spindles are operated on by a batch of 3 skilled labourers, instead of 4 labourers employed formerly.

The degree of twisting which is imparted to a thread has a great importance on its strength. The number of turns, *per inch*, is to get a particular quality of the thread. The quality of the yarn is, of course, dependent also on the quality of the raw fibre used. The finest quality of jute may produce yarn up to 20 counts. But the Indian mills spin up to 10 counts only; ordinarily for sacking purposes, yarns of 2 or 3 counts are used.

The *measurement of yarn* of jute is ordinarily made by *spindle* of 14,400 yards length of thread, which may be spun from 8 lbs. of jute for the purpose of warp thread, for weaving ordinary sacking cloth, equivalent to 2·14 counts. For spinning sacking weft thread, 12 to 13 lbs. of jute are required for one spindle. *Count* is a common term applied to the measurement of all sorts of thread. It is measured by a length of 840 yards for each count from the production of

1 lb. of fibre. The whole length of the thread from 1 lb. of fibre is thus divided by 840 yards, giving the *number of counts* of the fibre spun. If *one pound of fibre* produces 14,400 yards length of yarn, being divided by 840 yards, it would be called a yarn of 17·14 counts. The finer the fibre, the higher the counts obtained.

Winding

As soon as spun, the yarn intended for warp is rolled on *Spools* and that for weft on *Cop-Bobbins.* The thread of the spool (14,400 yards) is called *Roll* and that on a *Cop-Bobbin* (900 yards) is named *Cop-Bundle,* the operations being termed as *Roll-Winding* and *Cop-Winding,* respectively. The yarn of the spool is again rolled on *Beams,* a process called *Beaming.* This *Beam* supplies the warp-thread in the act of weaving, while the Cop-Bundle feeds the *Shuttle* which shoots the thread of the woof to and fro between the threads of warp when intercepted by *Cambs* (*Jhump,* a Bengalee word for jump.)

There is a large business in Calcutta in exports of *Twists* and *Yarn* in which the yarn containing single thread, or two or more threads being plied or twisted together according to requirements, is packed for shipment. So it is required to wind the yarn into skeins, cops or tubes, or in other shape. *Skeins* are made by leading the yarn from bobbins which are set on pegs to revolving reels and then removed. Skeins are loose in structure and are most convenient for dyeing. *Tubes* are large cylinders or cone-shaped yarn closely wound, while the *Cops* are small elongated cylinders of yarn used only in weaving.

Twine or string of various sizes and shapes is also exported. It may be either finished or unfinished. Twine is made up of a number of plies of yarn twisted together. The plied yarns may be converted, in turn, into still heavier cords by twisting several of them together. The *finished* twine or string is to be coated with thick paste of starch or other material scrubbed, dried and polished.

Reeling

The *Reeling-process* is resorted to when the yarn is subjected to bleaching, dyeing or woollenizing, for which *skeins* are made leading the yarn from a row of bobbins set on pegs to revolving *reels* of 4′–6″ in circumference, each. *Skeins* thus wound are loose in structure and they are most convenient for dyeing.

Dressing or Sizing

The operation of dressing the yarn with starch preparatory to weaving is not commonly in practice in the Indian mills. In this process, the yarn is treated with hot starch-paste and dried up before it is rolled on beams for weaving. The yarn is pulled through a trough containing hot starch-paste or any other glutinous substance and it is dried by being carried over a series of large copper drums heated from within and then wound upon a beam.

The starching action does not appear to be paying for such stuff as cheap sacking cloth. Complaints are however made that the Indian sacks retain dust which is not easily shaken off, which may be due to omitting the use of starch on the yarn.

CHAPTER V

Weaving

IN the Indian Mills, jute is chiefly manufactured into cloths known as *gunny* and *hessian*. For the manufacture of gunny, strong coarse jute is used, while hessian is woven from the finer quality, as canvas or wrapping cloth. Gunny means a strong coarse jute cloth or sacking cloth for making sacks, bags, etc., but both gunny and hessian are now called by the common name of *gunny* in the market, although the qualities are different. *Hessian* cloth is also used for making bags of superior quality. The coarse cloth and sacks are called *heavy goods*. Hessians are manufactured with warps of single thread and a single thread weft woven with two cambs. Sacking has three qualities, *viz.* :—

(*a*) Twill is made of double thread warps and single thread wefts. Three cambs are to be attached to a loom to weave a twill cloth.

(*b*) Heavy " C's " is woven with double thread warps and single thread wefts, adjusted with two cambs.

(*c*) Plain quality has single thread warps and single thread wefts woven with two cambs.

In weaving, the words *warp* and *weft* or *woof* are termed as *porter* and *shot* respectively. The thread extended lengthwise is called *warp*, and that which crosses the warp from selvage to selvage, actually woven, by means of a shuttle in the weaving, is called *weft*. Selected strong fibre is always used for warp and all inferior stuff for weft.

The terms *porter* and *shot* denote the *number of threads, per inch*, in warp and weft respectively, in a web. The standard width of hessian is 40 inches, and 8 oz. in weight per yard lengthwise, and known to be of 9 porters and 10 shots. This quality is called *light hessian*, while the *good hessian* is of 11 porters by 12 shots, 40 inches wide and $10\frac{1}{2}$ to 11 oz. in weight. There are 40″—$8\frac{1}{2}$ oz., 40″—7 oz., 40″—$7\frac{1}{2}$ oz., etc., manufactured, according to demand. In

weaving, shrinkage in finished goods, according to differences in qualities, is considered and made up. There is 4 to 8% of shrinkage in the length and about 4 per cent. in the breadth of a cloth.

Commonly a large business is done in hessians ranging from 6 to 14 oz. per yard, 40 inches wide. Of gunny bags produced in the mills in Calcutta, about 10 per cent. are made of hessian cloth and the rest of sacking cloth. Bags are commonly one yard lengthwise, requiring 2 yards cloth and 3 inches extra for hemming.

The hessians manufactured in the Calcutta mills are graded into four classes, A, B, C and D, according to the differences in the qualities of the products.

In weaving plain cloth alternate warp threads of the beam are so arranged as to pass through two frames called *harnesses* or cambs which lift and drop alternately. Thus one set of warp thread is up and another set down, intercepting the thread of the weft. Each time the cambs shift position, making an opening for shuttle to ply with cop yarn. There must be at least two cambs for two distinct sets of warp threads to form a web or cloth. More than two cambs are necessary when warp threads are intersected into more than two sets of warp threads required for weaving figures on the cloth. Three cambs are used for weaving twill cloth, while for figured fabrics, more than half-a-dozen cambs are wanted for separating the warp threads into sets. In the manufacture of fabrics of finer kinds, by power looms, Joseph Marie Jacquard devised an invaluable apparatus, about a century ago. It is called the *Jacquard-loom.*

Finishing

In finishing the cloth after weaving, it is damped by sprinkling of water through a pipe of the machine and pressed hard in a heated calender or roller machine to smooth it. Then it is measured and folded, called *lapping*, each fold being 2 yards long. Lastly it is baled. In regard to bags, it has similarly to undergo damping, calendering and measuring, and then it is cut into sizes as required, hemmed, sewn, and lastly baled for the market.

Ordinarily 2,000 yards of cloth (cuts of 100 yards each) or 400 bags in 16 bundles (1 bundle =25 bags) make a bale, measuring 10 c.ft. each, the density of a cubic foot being 56 pounds, or 2,800 lbs. for 50 c.ft. or a ton.

CHAPTER VI

Indian Mill Hands

IN the mill districts in Calcutta, there were over 262,000 labourers working during 1917-18. There has, of late, been labour unrest in India too. This difficulty, when it arises, should be tactfully dealt with. It must be remembered that the Indians are, in general, a sensitive people. An Indian, especially referring to the poorer classes, leads a very simple life. He lives on poor and cheap food, sleeps in huts, even under the shade of trees if required. Owing to this and other circumstances, the capitalists would suffer more from a strike than the poor strikers. If the authorities bear in mind that any kindness shown to them would hardly be forgotten, as is characteristic of an Indian, there would be no strike. If there be a strike in a mill, to avoid future trouble, kindness should always follow the strike. The Calcutta mills employed, on a daily average, about 325,000 persons, during three years from 1929 to 1931.

Labour scarcity may occur in the mill districts, at any time, should there be an outbreak of epidemic disease, for the labourers would flee without any consideration for money or comfort. The mills should therefore do all in their power to guard against epidemics.

Ordinarily, labour scarcity occurs in the mill area twice in a year, firstly, from the end of April to the middle of July, and secondly, from the end of November to the beginning of February, which are the sowing and harvesting seasons of paddy, respectively.

In the mill districts, there are principally four classes of labourers. The most important among them are the Bengalees who are intelligent, steady and fit for any skilled work. Next come the up-country men, including both Hindus and Mahomedans, who are hardworking hands, but rough in their dealings with others and slow to be persuaded by reasoning. They often indulge in drink, getting intoxicated and sometimes causing disturbances in the mill localities. The Uryas are a class of men who are intelligent and docile, but timid. There is another class of workers who will not stick anywhere. They will work for a few days in one mill and then go away to another. They are not

at all reliable and cannot be depended on. At the present time, Madrasee labour is also imported, while the Bengalee workers are getting smaller and smaller, forming only about 10 per cent. of the whole labour force employed in the mills. About 40 years or so earlier, the mill hands were all Bengalees. The wages earned by a labourer in the mills are not perhaps sufficiently attractive to a Bengalee.

Wages are commonly paid to a labourer on the following scales, as given in the Industrial Commission Report, 1918* :—

Nature of Work.		Amount per mensem.
Carders	(about 6% of the labour force)	Rs. 9
Drawers	,, 6 ,, ,,	———
Rovers	,, 7 ,, ,,	,, 12
Spinners	,, 14 ,, ,,	,, 15
Shifters	,, 14 ,, ,,	,, 11
Winders	,, 11 ,, ,,	,, 18
Beamers	,, 1 ,, ,,	,, 22
Weavers	,, 20 ,, ,,	,, 27
Mistries	,, 9 ,, ,,	,, 30
Coolies	,, 12 ,, ,,	,, 13

The following statement of wages compiled by the author, early in 1931, may be compared with that of 1918.

Wages of labourers for working 40 hours, in 4 days in a week, on contract system :—

Operators.	Weekly Amount (on average.)
Batchers	Rs. 2-10-6
Cutters	,, 2-12-0
Feeders	,, 2 - 8-0
Cleaners	,, 3 - 7-6
Rovers	,, 2-14-9
Shifters	,, 2-12-0
Rollers	,, 4-14-9
Spinners	,, 3-11-9
Twisters	,, 2-14-0
Reelers	,, 2-10-6
Beamers	,, 4 - 8-0
Weavers	,, 5 - 0-0
Coolies	,, 2 - 9-0
Sirdars	,, 5 - 0-0 to ,, 6 - 0-0

*During this period the mills were working, it is understood, for 54 hours a week.

For want of education, Indian labourers have little sense of self-respect. They would idle their time if not properly supervised. In the contract system of work, they are all right. Both for the mill people's efficiency and morals, the mills should arrange night schools for them and provide them with good dwelling barracks within the mill compounds. In addition to a course of primary education, a course of regular lectures on sanitation, duties, etc., should be introduced. Provision may also be made to afford occasional recreation to the workers. The mill authorities should establish co-operative stores to supply the labourers with necessary articles, at cost prices, in order to save them from the hand of unscrupulous *banias*. A provident or bonus fund is urgently required to be opened in each mill for the welfare of the mill people. The labourers might be helped from this fund at times of unforeseen difficulties and on retirement. Thus, in short, the efficiency of the Indian labour force can greatly be improved and the possibility of a strike would be minimised. Special care should be taken of women labourers at the time of child birth. The mills' authorities are requested to establish Maternity Homes and to employ women doctors.

We think it well to incorporate into this work, some of the most urgent recommendations relating to the improvement of labour employed in the jute mills, contained in the report of the Whitley Labour Commission. It may be mentioned that the Commission began work in October 1929 and completed it in March 1931. They are :—

(1) Whenever the scale of the factory permits, a labour officer be appointed directly under the Manager. His main functions should be in regard to engagements, dismissals and discharge. Wherever possible, an allowance should be given to the worker who goes on leave after approved service.

(2) The education of the industrial worker and that of his children should receive special attention of the mill authorities. The employers might provide buildings for schools with necessary equipments and render assistance in other ways.

(3) The weekly working hours should not exceed 54 hours, limiting daily working hours to 10. For women the rest period should include the hours between 10 P.M. and 5 A.M. and for children the rest period 7 P.M. and 5-30 A.M. and the daily working hours limited to 7½ hours. Persons between the ages of 15 and 16 years should not be employed as adults without medical certificates of physical fitness.

(4) Every factory should be compelled to maintain separate and sufficient latrine accommodation and adequate staff to keep them clean. The provision of water and places for washing should be obligatory.

(5) A provident fund might be opened for those employees who would draw not less than Rs. 20 per mensem ; membership being optional and eligible after one year's continuous service.

(6) All classes of workers should enter into a simple service agreement providing for (*a*) Probationary period of 12 months ; (*b*) Confirmation after 12 months' approved continuous service ; (*c*) confirmed service terminable on one month's notice.

(7) A more generous policy in respect of recognition of trade unions is desirable.

(8) Every effort should be made to put into operation, a policy of standardized wages, both for time and piece workers.

(9) Fines.—

(*a*) Children should be exempted from fines. (*b*) The payment of fines should not be spread over more than one month from the date on which it is imposed. (*c*) The maximum amount deducted, in fines, should not exceed, in any month, half an anna in the rupee of the worker's earnings. (*d*) The sums received from fines should be credited to a beneficial purpose to the employees. (*e*) Employers should be required to post notices specifying the acts or omissions in respect of which a fine might be imposed.

(10) Policy should be adopted of restricting the facilities for the sale of liquor.

(11) Authorities should construct sanitary markets in all urban and industrial areas.

(12) Provision of proper housing for workers should be statutory obligation.

CHAPTER VII

Rural Economic Condition and the Co-operative Societies

AGRICULTURE is the only industry in India on which as much as 75 per cent. of the people depend for their livelihood. There is no other civilised country where the people are dependent on agriculture, in such a proportion. Natural proportions between agricultural and non-agricultural people should be half and half, one-half supplying the food grains and raw materials, and the other half the essential manufactured articles such as clothing, implements, etc. Due to the abnormal adjustment of the agricultural and the non-agricultural population, nearly the whole country is affected by even a partial failure of main food crops, in a bad agricultural year. Even in normal conditions, the raiyat lives on a very poor diet consisting of two meals daily of 12 ozs. of rice each time, with some pot herb and some fish if available, and also a morning collation of half a pound of soft rice containing gruel or the same quantity of cold rice and water, the whole costing 10 to 12 pice per head. Ordinarily he cannot save food grains for more than 6 months of a year. For the next 6 months' expenses, he has to borrow. It would, however, be of great benefit to the country, if the present indebtedness of the peasant classes is relieved to any extent by offering loans at reasonably small interest, not exceeding Rs. 12 per cent. per annum.

The present rates of interest, from Rs. 3-2 to Rs. 6-4 per cent. per month, are simply ruinous. If a raiyat incurs a debt at such high rates of interest, he can hardly get free from the money-lenders. He cannot save anything, even from a bumper crop, after making payments to the landlords, money-lenders, etc. Six months will not lapse before he will have to go to the money-lender again for a fresh loan.

It is true that the money-lender cannot realise such high interest if he go to the Civil Court, but the debtors

themselves do not generally intend to be dragged to the Civil Court, fearing, in such case, they will lose their credit, in future, and none will help them with any further loan, in time of need.

What is now required is to relieve these toiling masses by establishing fully organised Co-operative Societies, in every part of the country. In the whole of Bengal there are only 15 thousand societies, urban and rural, with about 5 lakhs of members who are eligible to transact business with such societies. The capital raised amounts to about Rs. 122 lakhs, of which the Government has contributed more or less half a lakh only. There will be no dearth of capital, if there is no mismanagement and misapprehension. Unfortunately the cultivating classes are mostly illiterate. There would be no success, unless the educated *bhadralok* classes are interested in the affair.

The defects of the Co-operative Societies, for which the institutions are not quite popular, should be removed. The societies, under the present system, lend money to the members to be realised after a certain period. The members may be permitted to pay up the debt, if possible, on any date before the definite time. The rate of interest should be reduced from Rs. 15-4 to Rs. 12 per cent. per annum, as the maximum. No money is advanced to a member, on security of his produce. The societies should give all facilities to their members by arranging for the sale of their crops on the most favourable terms. The progressive societies should keep agricultural implements and bullocks to be worked in the fields of the raiyats when requisitioned. Each society should keep a breeding bull of proper type. The societies should arrange to cultivate or stock fodder, to be supplied to the members when fodder gets scarce. The societies should also stock common seeds and manures, to be supplied to the members at sowing time. A fully organised society is expected to render assistance to the villagers in all matters relating to agricultural economy and finance.

The seriousness of the present depression in the jute market seems to have at last attracted the attention of the public as well as that of the Government. Various suggestions are appearing in the press daily. It is not practicable, as frequently suggested, for an ordinary raiyat to withhold his stock, till a reasonable price is offered for the stuff.

Most of the raiyats are poor. They cannot afford to stock their food stuffs for more than 6 months since the time

of harvest of paddy, and the jute season is the time when they are urgently in need of ready money for their own living and for paying a *kist* to the Zemindars, and also dues to the money-lenders.

The question now arises, as to the acreage under jute which should be maintained, without harming the manufacturing industry. In our opinion the cultivation and the industry should meet, looking after each other's interest, and some standard prices for different qualities ascertained. If the cultivation does not supply raw material in sufficient quantity, the industry must suffer, or fail altogether. If the industry fails, the cultivation also dies in due course. It is advisable for the well-organised industry to help the disorganised cultivation, to be in a thriving condition. From our long experience, we are in a position to say that the cultivation of jute has in recent years made praiseworthy improvements in the qualities. Good qualities of fibre are now grown in the area where very bad qualities were produced before. The present slump is likely to induce the growers to neglect the crop, to its deterioration.

The normal demand for the jute crop is estimated at about 100 lakh bales, of which 55 lakh bales are to be consumed in the Indian mills, 5 lakh bales for home consumption of the raiyats, and 40 lakh bales are for exports out of India. Bengal should grow 85 lakh bales, 15 lakh bales being estimated to be the production of other provinces. It is required for Bengal to grow jute on 25 lakh acres to yield nearly 85 lakh bales. For the present Bengal is advised to grow jute on 20 lakh acres only in adjustment of the excess stock.

Unfailing effort is needed to organise the peasantry of this country into co-operative bodies, in order to give effect to any scheme of improvement of the cultivators. The Co-operative Department is expected to do this work successfully. The Department should establish selling marts at different centres, where they could store the raiyats' crops after making some advances to them, and sell the different qualities afterwards at the most advantageous prices. The sale proceeds would be distributed after deducting the advances paid and reasonable commissions charged for the expenses and the interest on the capital invested. This scheme must succeed if properly worked out, under proper supervision, and it is the only way which will strengthen the position of the cultivators, who are at present helpless and dependent upon the money-lenders.

PART VI—MISCELLANY

CHAPTER I

Substitutes for Jute

JUTE is the most widely used material in the world for making sacks for packing cotton, sugar, coffee, grain, wool, etc. India exports, on an average, over a thousand million yards of jute cloth each year, and about 500 million bags made of the material. Moreover, India exports over five million pounds of jute yarn, not to speak of the raw jute, amounting to over 600 thousand tons, for manufacture elsewhere. Up to the pre-war period, there had apparently been little effort made to find a substitute for jute, as its cost was regarded as fairly cheap. But during the war and the post-war period, the price of jute and the jute products had gone up so high in other countries, that substitutes were being sought. A plant known as *Malva*, heretofore considered a weed, that grows wild in many parts of the island of Cuba, was tested and found to be promising. There are not less than eleven species of this plant in the island, but the one that is considered the best suited for the production of textile fibre is known as *Malva blanca*, and it has, of late, been cultivated to some extent. *Malva* is said to grow sometimes to a height of about twenty feet, but on an average from six to ten feet. The stalk of the plant ranges from half an inch in diameter to three times that size. The fibre is contained in the bark. The young plant contains a single layer, while older ones as many as eight layers. This fibre is said to be equal in strength to the Indian jute, but finer and resembling flax, and it is therefore believed that it will be found useful for rope making, as well as for bag fabric. In Brazil also, a plant called *Paco-paco*, probably of the same family, grows wild, from which fibre is extracted to a considerable extent.

Within recent years, in the United States of America, a paper bag called " *Multy-wall* paper bag " is manufactured on a large scale. It is made up of five thicknesses

of paper pasted with glue, sometimes with thread sewn seam and sometimes bottoms, and both tops and bottoms are folded over several times, then fastened by wire staples driven through by a machine which clinches them. The five-ply paper bag has already gained a strong hold on the cement trade in America, displacing the jute bags, formerly imported into that country for this purpose. The *Multy-wall* paper bag was sold at 4·8 cents each, whereas the price of the jute cement bag of the same capacity containing 100 lbs. of cement, was 10·2 cents each, during the months of June-July, 1929. Although it is not claimed that the paper bag is equal to the jute bag in quality and in handling, but due to its cheap price, it has become a very strong competitor of the jute bag. They are trying to manufacture a heavier paper bag of twisted paper yarn of greater wearing capacity, suitable for packing sugar and flour. The jute manufacturers should, however, take courage, remembering the fact that there is an immense quantity of cheap raw material, in India, with which they could produce cheap jute bags to compete favourably with the paper-bags. Jute bag is manufactured direct from the fibre, while for the other, it is required to manufacture pulp at first from the wood, then paper, and lastly bags. It is not understood how paper bags could be made cheaper than the jute bags. Mills should cut down expenses and increase efficiency, and thereby produce cheaper goods.

It is useless to issue a warning to the Indian jute growers, who are dumb people without education and capital. But we cannot do better than request the authorities of the Indian jute mills who have freely invested capital in this country, in the jute industry, and the Bengal Zemindars who realize easily the rents from their tenants, on account of the ready sale of the stuff, to take joint action against any sort of possible danger to the crop and the industry.

Hibiscus cannabinus and *Crotalaria juncea* are the next most important fibrous crops of India, which we will discuss in detail. Brief accounts of some other different bast fibres are also given by way of comparison with jute.

Hibiscus cannabinus

(*Mestha pat*)

It is called *Mestha pat* or *Mesot* in Bengal, *Patua, Kudrum, Channa* and *Amla patua* in Bihar and Chota Nagpur, Deccan *Hemp, Ambari* or *Ambadi* in Bombay, and

Gogu and *Pulicchai* in Madras. *Bimilipatam* jute is another name of this fibre. In some districts in East Bengal, both jute and *mestha* are known by the name of *koshta.*

It is said to be indigenous to Africa and introduced into India. It is now extensively cultivated in the Central Provinces, Madras (about 64,000 acres during 1929-30) and in Bombay, especially in the Deccan and Karnatic.

The *Mestha pat* is the most accommodating fibrous crop known. It grows on all kinds of soils and lands—alluvial or laterite, high or low. Some years ago, it was largely cultivated in Eastern Bengal, but of late, jute has completely taken its place. There is still a limited cultivation of this crop on the *bheel* lands which are considered too low to be suitable for jute. It is commonly sown as a hedge crop in Bihar.

On the *bheel* lands, in Bengal, it is sown in February and March, and harvested in July and August. In Bihar and Chota Nagpur, it is generally sown in June and July, and harvested in September and October. It thrives pretty well, in other seasons, if the soil has sufficient moisture. Thirty seers of seed are ordinarily required for sowing one acre of land, if it is sown by itself. *Mestha* is commonly sown as a mixed crop with paddy in East Bengal, and with sugar-cane, *rahar* and cotton in Bihar. The quantity of the seed is, of course, to be lessened then. The plants are so thinned as to allow a space of about 9 inches around each plant. The flowering stage is considered the best time for harvesting. It takes about four months from the time of sowing to the time of cutting. In East Bengal, the plants which grow in the *bheel* lands are pulled up and steeped as jute; and the method of extracting fibre is the same as that of jute. In Bihar and Chota Nagpur, the bundles of plants, when they are ready for stripping, are taken away from the steep and dried; afterwards the fibre is separated from the stalks. Fifteen maunds of fibre, per acre, are considered to be a good yield, in East Bengal, and ten maunds in Bihar and Chota Nagpur, if the crop is sown by itself.

The fibre of *Mestha* is bright and glossy, but coarse, harsh and brittle. On the whole, it is inferior to jute both in quality and yield. *Mestha pat* is commonly used in cordage and sacking.

Crotalaria juncea.

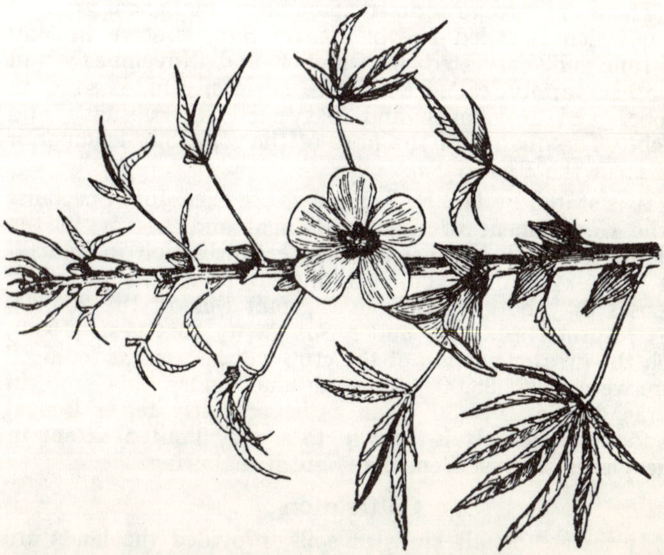

Hibiscus cannabinus.

Crotalaria juncea

(*San pat*)

There are a series of vernacular names for this plant, of which *San*, *Sanai* and *Sani* are common. In Madras, it is called *Shanal* or *Janumu*. It is commercially called *San* or *Sunn* Hemp, Indian Hemp, Bombay Hemp, Jubbalpur Hemp, False Hemp, and also Travancore Flax (*vide*

Sir George Watt's " The Commercial Products of India "). According to Mollison also, the *San* and Bombay Hemp are the same. But, from the Season and Crop Reports of the Madras Presidency, they appear to be different. Regarding this confusion, we drew the attention of the Director of the Department of Agriculture of the Province. He has been pleased to inform us that the Sunn Hemp and the Bombay Hemp are different crops, *Crotalaria juncea* and *Hibiscus cannabinus*, respectively. As the terms Sunn Hemp and Bombay Hemp are used to denote the same fibre in other Provinces, and also in the trade, it would be well to adopt a common name for it, all over the country. Common local name is *San pat* in Bengal and *Sanai* in Bihar.

It belongs to the Pea family of plants (*Leguminosæ*), with narrow leaves 2″—3″ long and bright yellow flowers.

There are two varieties of *San* hemp in Bengal ; one of them, which is called *Ful* or *Bhadoi San*, is sown in May and June and harvested in October and November ; and the other variety, called *Gour San* or *Rabi San*, is sown in October and November and harvested in February and March. The quality of the fibre of the *Rabi San* is superior to that of the *Bhadoi San*.

It is stated by Mr. R. S. Finlow, the then Jute Specialist to the Government of Eastern Bengal and Assam, in the Bulletin on " the Extension of Jute Cultivation in India," that " the extent of the cultivation (*San*) in Madras for fodder purposes may be gauged from the fact that of the 205,000 acres returned as being under *San* hemp this year (1906), much the greater portion of the crop will be used as fodder." There were about 88,000 acres of land under this crop in Madras, during 1929-30. *San* is never cultivated in Bengal as a fodder crop. It is grown to a very limited extent in some places in East Bengal for green manuring.

Cultivation

It thrives on all kinds of soils, provided the lands are well drained and retentive of moisture, and above the inundation level.

The seed is sown broadcast thickly—2 maunds being required for one acre of land. The sowing must be very thick, otherwise the plants will grow bushy, causing the extraction of fibre extremely difficult. The *rabi* variety may be sown in rotation with jute or *aus* paddy in the same year. This variety cannot be grown in Chota Nagpur and

Bihar, excepting Purnea District, for want of steeping water and for deficiency of moisture in the soils.

The plants are either pulled up or cut to the ground. The best fibre is obtained when it is cut at the flowering stage. Generally the crop is harvested when the seed ripens, in order to get a heavier yield of fibre, and also the seed.

In East Bengal, after the plants are cut or pulled up, they are spread over the fields for withering. On the second day, the plants are tied into bundles and submerged under water, applying some artificial weight (commonly sods of earth) over them. There is another process of steeping, in which the bundles are placed upright in water about 2 feet deep for one day, submerging them completely on the next day. Retting generally finishes in 3 or 4 days. The plants are stripped immediately after retting is complete, or else the fibre will be totally lost. It is stripped in the same way as jute in Orissa, that is, handfuls of retted plants are beaten over water to remove gummy matters from the fibre, which is then dried and separated. The great obstruction in the way of its extended cultivation is the difficulty in stripping. The yield of fibre is between 9 and 12 maunds per acre.

The *San* fibre is decidedly stronger than jute, specially when wet, but the yield is much less, while the cost for stripping is much more. In India, the fibre is very largely used for making fishing nets, but to a limited extent for weaving rugs. Excepting *Rhea* (*Kunkura*), which is very expensive, there is no other fibre in India which can replace *San* in the use of fishing nets. Wet or damp jute rots quickly, and, therefore, it is useless for fishing nets. Cotton yarn is now commonly used for this purpose, for it is cheaper than *San* yarn. An everyday used fishing net made of *San* lasts for about three years, while a cotton-net gets impaired after it is used for two years, at the longest.

Area

The Government Reports do not show separately the area under Sunn hemp and Hibiscus. The area under other fibres, excepting cotton and jute, as given in the Statistics, represents *Sunn hemp* and *Mestha* together. In Bengal it is Sunn hemp alone, on account of Mestha fibre being included under jute. Considering the importance of these fibres, especially that of Sunn hemp, in commerce, being required for carpet and twine manufactures, *other fibres* should be reported separately.

From the Statistics published by the Provincial Departments of Agriculture, other fibrous crops cover a considerable area, in each Province. The figures, as reported by the Departments, are quoted as follows :—

Season 1929-30.

Area under other fibres other than Cotton and Jute.

Bengal ... ... 64,000 acres
Pabna District, being the most important ... ... 30,000 ,,
Mymensingh ... ... 20,000 ,,
Burdwan District ... ... 5,000 ,,

U. P. and Oudh ... ... 184,000 acres

The following are the important districts :—

Benares ... ... 22,000 acres
Jaunpur ... ... 20,000 ,,
Partabgarh ... ... 14,000 ,,
Bareilly ... ... 13,000 ,,
Allahabad ... ... 11,000 ,,
Pilibhit ... ... 10,000 ,,

Punjab ... ... 41,000 acres

Central Provinces ... ... 86,000 acres
of which
Berar Division ... ... 45,000 acres
Jubbalpur Division ... 20,000 ,,

Bombay Presidency ... 84,000 acres
of which
Deccan ... ... 44,000 acres
Karnatic Division ... ... 18,000 ,,

MADRAS

Total ...	... 163,288	acres
of which under Sunn hemp ...	88,421	„
under Bombay hemp (said to be Hibiscus) ...	63,838	„
and under Jute ...	... 856	„
Vizagapatam total	... 55,511	acres
of which under Bombay hemp	45,754	„
East Godavari total	... 12,941	acres
under Sunn hemp	... 12,395	„
Kistna total ...	... 18,948	acres
under Sunn hemp	... 15,937	„
Guntur total ...	... 62,459	acres
under Sunn hemp	... 50,858	„
„ Bombay hemp	... 11,594	„

Trade

The following statement shows the importance of Sunn hemp as a commercial fibre of India :—

Exports of raw fibre.

Year.	Quantity. Cwts.	Value. Rs.
1930-31 ...	... 293,151	39,30,344
1929-30 ...	... 434,959	68,33,000
1928-29 ...	... 561,097	77,75,000

Belgium is the principal buyer. About half the quantity of the exports goes to Belgium. The United Kingdom, Germany and France are the next important buyers.

Of the exports, Bengal shares about 75 per cent., Bombay nearly 25 per cent., and Madras about 3 per cent. The Indian hemp fibre contains excess of dust, dirt and sticks, and it is most carelessly and irregularly packed. The value per ton of Indian hemp in Europe

is varying between £17 and £37 owing to its containing dirt in different proportions. The fibre for exports should be packed after it is subjected to a hand-hackling or combing process, in which dirt and short fibres are mostly removed. Owing to its containing too much dust, the working people in Europe have been offering resistance to work with it and consequently Indian hemp is being replaced by Sisal hemp.

From the examination of the samples of the Indian San hemp fibre by the Imperial Institute, London, it would appear that the quality of the fibre is not far inferior to that of the true hemp fibres of Europe. But there is a prejudice against the Indian hemp in the commercial circles in Europe, not due to any material defect in the fibre itself, but to the careless conditions under which it is prepared and marketed, as already noticed. The results of relative strengths of different qualities of fibres are as follow :—

Breaking strain in lbs. on average.

Before immersion :—

Indian hemp (San)	91·8
Russian hemp (True-hemp)	116·6
Italian hemp (True hemp)	126·0

After being exposed in sea water :—

	After 1 month.	After 2 months.	After 3 months.
Indian hemp ...	66·1	35·9	3·2
Russian hemp ...	81·9	31·4	4·7
Italian hemp ...	96·4	50·5	7·1

After being exposed in fresh water :—

	After 40 days.	After 80 days.	After 160 days.	After 282 days.
Indian hemp	83	77·1	72·1	70·3
Russian hemp	94	88·0	85·2	70·1
Italian hemp	103	101·9	102·0	92·2

The Russian and the Italian hemp is obtained from *Cannabis Sativa*, the true hemp plant which is cultivated in Bengal as a narcotic drug only. In some localities in the North-West Himalaya, it is cultivated for fibre chiefly for local consumption.

From Calcutta, the Sunn hemp is exported under the following three grades :—

(*a*) Benares quality.

(*b*) Raigarh quality.

(*c*) Bengal Proper quality.

Roughly about 100,000 bales of 350 lbs. each is annually exported, during the season, from October to May. In a normal year, U. P. supplies about 85 per cent., Raigarh (C. P.) 10 per cent., and Bengal Proper 5 per cent. The Bengal fibre is the best in quality, Benares quality is the second, while the Raigarh grade is much inferior.

Cannabis sativa

(*The True Hemp*)

The plant is commonly known by the name of *bhanga* and is found in a wild state all over the northern parts of India, extending up to the Himalayan altitude of about 7,000 feet. Although it has been used in India as a fibre plant and cultivated from very early times, it is not, now-a-days, cultivated in the plains of India, except to a very small extent in Sind for fibre and in Bengal for its narcotic property, when it is called *ganja*. As a source of fibre, it is chiefly grown in the North-West Himalaya, in the districts of Garhwal and Kumaon, and also in Kashmere and Nepal States. The fibre is principally used for the manufacture of rope and string, and also for coarse cloth commonly used as garments in Garhwal. Russia is the most important country for this hemp fibre. It is also cultivated in China and Italy.

Mode of Cultivation in India

The seed is sown on well cultivated lands, at the rate of 60 lbs. to an acre, from March to June. By September or November the crop is ripe for harvest, when it attains to a height of 10 to 14 feet. During the growth of the plants, the land is twice hoed and the plants are thinned out like those of jute. The male plants yield fibre of better quality. The stems, at the flowering stage, are cut and spread on the ground for 24 hours to shed the leaves. Afterwards the stems are retted and the fibre separated. An acre of land yields about 10 maunds of fibre.

For the narcotic hemp, the male plants are carefully removed from the fields, as early as possible, before the flowering buds develop. The mature female plants, after a resinous substance appears on the inflorescences, are cut down and cured and sold as *ganja*, which is either used for medicinal purposes or as an intoxicating smoke by the people of lower standard of habits. Excessive use of the hemp, in any form, weakens the mind and may even lead to insanity. It cannot be cultivated in British India, without license from the Government. The dried leaves of the plants (specially female), not so cured as *ganja*, are also taken after being pounded and diluted with water, milk and sugar for intoxication of a milder form.

Cannabis sativa in China

The true hemp is one of the oldest known textile fibres. It is on record that this hemp plant, known by the name of *ma* (now-a-days a generic term for fibre yielding plants), was cultivated in China for fibre 27 centuries before the Christian era. For its strength and durability, it has been held in esteem, since the earliest time. Until less than a century ago, this hemp and the flax were the principal vegetable fibres, the flax being used for fine linens, laces and embroideries, while the hemp for ropes, cables and sails. Canvas is the Arabic name for the hemp, from which the commercial name of canvas has been derived. The true hemp for marine cordage has been superseded by Manila hemp, on account of its superiority in resistance to injury from salt water.

The annual production of hemp in China is about 32,000,000 lbs. (205,000 bales), while the production of flax in China is estimated at 267,000 bales.

Abuliton Fibre of China

Abuliton fibre (*Abuliton avecennæ*) is extensively cultivated in China. Sometimes it attains a height of 15 to 20 feet. True hemp fibre is stronger but shorter and darker in colour and more expensive. Both sown in April and harvested in August. Both the fibres are used principally in rope and cordage, and also rugs, twine, canvas, bags and carpets. Sack cloth is chiefly manufactured from the true hemp. Fishing nets are commonly prepared from abuliton fibre for its possessing greater water-resisting properties. The plants of both are pulled out of the ground

and seeds removed. The stems are retted for about 10 days, until the fibre no longer sticks to the woody part. The plants are then taken out of the water to dry. The dried plants are pounded to separate the fibre from the woody stems.

Ramie Fibre (Bœhmeria tenacissima) or Rhea (Bœhmeria nivea)

It is called *kunkura* in Rangpur and Jalpaiguri. It is also sometimes called *kurkunda* in the latter place. In Bogra it is known by the name of *kund*.

The undersurface of the leaf is white. *Ramie* and *Rhea* are the two forms of the same species and the fibres of both the forms are identical. China is the only country which grows the fibre on a commercial scale. It does not pay the cost for production in this country. The Chinese name of the plant is Hsien-ma.

Soil for the cultivation of Ramie.—Rich, light, sandy loam exposed to the sun (not shady), containing sufficient moisture, is considered by the cultivators to be best suited to the cultivation of *Ramie*. Homestead land is generally chosen for it. High lands which are well-drained, that is, where water does not stand, are suitable for the cultivation.

Preparation of the soil.—The land, rich or poor, intended for the cultivation of *Rhea*, is liberally manured with cowdung. Ashes are scarcely used as manure; on the contrary, it is said that ash manure injures the crop. The Rhea plantation requires deep ploughing or spading.

Planting and treatment.—It is propagated from root-cuttings and sometimes from stem-cuttings of matured plants. Cuttings are made 6 to 9 inches in length and planted under 3 to 4 inches of soil, 1 to 3 feet (commonly $1\frac{1}{2}$ feet) apart from each other. There are two seasons for the plantation of *Rhea*, *viz.*, April and May, *i.e.*, before the commencement of the rainy season; and secondly, September and October, *i.e.*, at the end of the rainy season. The majority of the cultivators are in favour of plantation before the beginning of the rainy season. But the cultivators of Barabari, in Rangpur, are in favour of plantation at the end of the rainy season. Those who prefer plantation in April and May say that they obtain cuttings within a short time, that is, three cuttings in the first year, the first cutting being obtained in July. And those who prefer plantation

in September and October say that they obtain better crops at four cuttings, the first cutting being in June. They also say that the plants, from the plantation before the rainy season, do not develop enough within so short a time that they can yield an average crop.

Rhea fields are weeded and hoed after each cutting. In the second and in the subsequent years cowdung and earth are applied to the fields in the cold season. Cowdung is followed by earthing to the depth of 3 to 4 inches.

Rhea is generally cultivated by the fishermen in Bengal.

Cutting.—Plants are cut down before they run to flower and before they bring forth branches. When the leaves of the plants within about 1½ feet from the ground fall down, the plants are considered to be fit for cutting. Considerable experience is necessary to decide the right time for cutting. Two to four cuttings (sometimes five) are obtained in a year. Plants are cut down in fine weather, so as to dry them considerably within 24 hours from the time of cutting. The plants are cut at the bottom about 2 inches above the ground. Poor soil does not yield more than two cuttings a year. On rich soil the plants grow rapidly to the height of about 5 feet in a month and become fit for cutting. The cuttings obtained from rich soils, if planted in September and October, are as follows :—

1st cutting in May (the worst).
2nd ,, in June (the best).
3rd ,, in July.
4th ,, in August or in the first part of September.

Supposing that a plot would yield 32 seers of fibre at four cuttings a year, 2 seers are obtained at the first cutting, 13 seers at the second, 10 seers at the third and 7 seers at the fourth. A well-cared-for plantation may yield fibre for several years, from the same field.

In the cold season, the cultivators do not cut the plants for fibre. In some places one cutting for the fibre is obtained in this season; but this fibre is the worst possible in quality. In some parts of Jalpaiguri, the plants are cut down once or twice, in the cold season, and thrown away for the sake of good crops in the next rainy season. They say that the fields do not give a good yield, both in quality and quantity, if the plants are allowed to stand on during the cold season.

Separation of fibre.—After the plants are cut down, they are scraped with a knife or a scraper made of bamboo. This operation is generally done with the bamboo scraper.

Care is taken to scrape all the plants cut down within a short time, not exceeding 24 hours. Then the stalks are kept in the sun for 4 to 10 days, in other words, until they are quite dry. The dry stems are broken into pieces and then the fibre is separated from the stalks. The act of separating the fibre is the most troublesome and expensive piece of work. Fibre is often damaged if the sky continues to be cloudy for some days together.

Dyeing.—It takes dyeing readily. In the Bogra district the *Rhea* fibre is dyed with turmeric powder in the following process :—Water is boiled with turmeric powder and the fibre is dipped into it. After a few minutes it is taken out and dried in the sun. This operation also makes the fibre softer. In one village of Bogra the *Rhea* thread is boiled in rice-washing water for about half an hour. It is said that this also makes the fibre softer. In some places of Bogra the nets made of *Rhea* fibre are dyed with a decoction of the bark of *Odina Wodier* (*Jiol*) to make them stronger.

Uses.—Rhea fibre is used in Northern Bengal for fishing nets and lines. In fresh water, it is said that the *Rhea* fibre is the strongest of all the fibres, including silk. So the fishermen purchase it at high prices. It is the strongest and finest bast fibre known. It looks more like wool than plant fibre. Tests of strengths of the following noted fibres, after being degummed, show interesting results :—

Ramie	100	units
Cannabis hemp	36	,,
Flax	25	,,
Silk	13	,,
Cotton	12	,,

In China and Japan cloths of superior and fine qualities are woven with this fibre.

Outturn and the cost of production.—Rhea is not so largely cultivated in this country that we may give reliable information about the outturn. The Bengal production of fibre, per acre, may fairly be estimated at between 500 lbs. and 600 lbs. In China, where it is more carefully cultivated, it yields 850 lbs.

A plantation is liable to attack by a caterpillar which eats the roots of the plants. This caterpillar is called by the cultivators of Rangpur *malpoka*. The *malpoka* causes such serious damage to the plantation that the plants wither away within a year or two, when the roots of the

fields are removed and transplanted. The new plantation, it is said, should get on well for three or four years.

The fibre is sold locally at about Re. 1 per seer (2 lbs.), but its commercial value hardly exceeds £50 a ton or about 6 annas a pound.

China's production.—China produces annually about 26,000 tons of fibre, of which about 20,000 tons are exported. More than half of the exports go to Japan. Germany and Great Britain are the next important buyers.

Sisal Hemp or Aloe Fibre.

Aloe Fibre ought to be one of the important fibrous crops of India, where it grows in dry and rocky waste lands. The species of Agave are indigenous to tropical and temperate South America, Mexico and Southern States of U. S. A., from where they have been introduced into other countries. True Sisal Hemp (*Agave sisalana*) came to India between 1885 and 1892, and was successfully cultivated in different parts of the country. But the spread of its cultivation in this country has been checked for the reason that the preparation of the fibre is done by hand-scraping and the acrid juices released cause painful blisters on the skin of the operators.

The fibres, which are produced in India, are sometimes pronounced good, fetching favourable prices, and sometimes inferior. The difference in the quality of the fibres is due to the fact that it is harvested and prepared in the right methods in some places, while it is not so in others. Old leaves yield coarse, hard and brittle fibre. The best time for cutting is when the leaves rest at right angles or at half right angles to the stem. Fermentation of leaves by immersion in water weakens the fibre. Green leaves, after cutting from the stock, should be scraped by hand or machine, and then washed and dried. From October to June is the proper season for harvest.

Plants are propagated by suckers from the base of the plant. Bulbils from the flowering spike are also used. Seeds take a long time to get plants ready for transplantation. Young plants are raised, at first, in a nursery, until they are 8 to 12 inches high, and then transplanted 6 to 8 feet from one another just before the rains set in, in rows 12 feet apart, to give roughly 650 plants to an acre. The plants come into full bearing when they are in the 4th year and continue to yield fibre till about the 12th year, when inflorescence or pole appears.

1,000 lbs. of cleaned fibre, per acre, more or less are annually obtained. One plant yields roughly 20 to 30 leaves, sometimes more, yearly, at the rate of about 50 lbs. of fibre from each 1,000 leaves.

Yucatan produces the largest quantity of Sisal Hemp (about 400,000 ·bales), for the world's consumption. Bahama, Mexico and Africa are also prominent supplying countries. Bombay exports some quantity of Aloe Fibre, estimated to be about 10,000 bales, annually. It is sold at about £20 per ton. The fibre is chiefly used for ropes, cordage, carpets, mattings and brush-making.

The importance of Sisal Hemp, as a commercial fibre, may be estimated from its consumption of 85,661 tons in the United States of America alone, during 1930, when the same country imported 64,052 tons of Manila Hemp and 61,025 tons of jute fibre. During that year, Sisal hemp was sold there at 6·2 cents per lb., and jute at 5·0 cents, or a little less.

Flax and Sida

Although flax was one of the oldest cultivated fibrous plants of India, it has been abandoned since a long time. Attempts for its revival in several places in India, by the British Government, have not been fruitful; sometimes the cost of seeds alone was not recovered.

Sida, a perennial wild plant of India of *Malvaceæ* order, which yields a fine and strong fibre resembling flax, was recently cultivated, but the results were not encouraging, from a financial standpoint. The plants are branchy, therefore extraction of fibre is most costly. Experimental plots gave 5 to 6 maunds of fibre per acre. Sida thrives during the rains, while flax is a cold weather crop.

Manila Hemp or Abaca Fibre

(*Musa textiles*)

Manila Hemp is produced from a plant which closely resembles the banana plant. The Philippine Islands supply the world's requirements. The approximate annual consumption of this fibre, in the United States alone, is roughly 150,000,000 lbs. It has been introduced lately in the Dutch East Indies. It grows from old stocks from generation to generation.

The Abaca fibre is chiefly used for the manufacture of superior grades of cordage. Its remarkable strength, elasticity and resistance to the action of salt water make it a most particularly suitable material for marine cordage. Over 200,000 bales of fibre are annually produced in the Philippine Islands. The United States and Great Britain are the principal buyers.

Fibre is obtained from leaf-sheaths. The outer sheaths give stronger but coarser fibre than the inner ones. The pulp and watery particles are removed by a blunt knife. A fibre cleaning machine has lately been invented in America which gives most satisfactory results. This machine-cleaned fibre is much superior to what is hand-cleaned. Abaca fibre is collected from flowering plants.

The Indian plantain fibres which we had collected on some occasions were not satisfactory both in yield and quality. Of the Indian plantain plants, a form which is very strong and tall, yielding fruits full of seeds known as *bicha kala*, would prove to be more suitable for fibre.

Manila Hemp could be successfully introduced in the moist hilly regions up to about 500 feet elevation, in India, where labour may be easily available.

CHAPTER II

Introduction of Jute into other Countries

UP to the present, attempts to introduce jute into other countries have not been successful. Repeated trials were given to it in Brazil, Egypt, Java, Philippine Islands and other places. It did not succeed even in Burma. The failures must have been chiefly due to unsuitable conditions of the lands which were selected for the crop.

Jute is said to be cultivated to a little extent in the plains of Ning-po, Tientsin, Mukden and Kirin in China, producing annually, on a rough estimate, 35,000 bales or 14,000,000 lbs. of fibre. It does not, however, appear to be an article of much importance in the trade of that country, as large quantities of Indian goods are imported by it every year. There was a demand of about 6,000 tons, in 1929-30, from China, for Indian raw jute.

Vigorous efforts were made by the Republic of Brazil to introduce the cultivation of jute. There is a considerable demand in Brazil for gunny bags of a superior quality for export of coffee. In the State of Sao-Paulo, there are about ten mills which manufacture packing bags of special sizes out of an indigenous fibre called *paco-paco* (probably of *Malvaceæ* order), which is much inferior to jute in quality. *Paco-paco* grows wild in some parts of Brazil. The mill-owners of Brazil import finer qualities of raw jute—about 15 thousand tons annually—from India, for mixing with *paco-paco* for the manufacture of bags. Sometimes they import Indian twist and yarn of good qualities for the purpose. Considering the heavy charges incurred for freight and the delay in transhipment, it was resolved to introduce the jute crop into their country. The crop which was under experimental cultivation succeeded in the plains of Sao Paulo. This area is watered by the tributaries to the river Parana, and its climatic conditions resemble those of India for a certain period of a year. The Amazon region, which is a vast country not yet reclaimed and peopled, appears

also likely to be suitable for the crop. But though the crop was a success, cultivation could not be introduced there, on account of difficulties in the extraction of the fibre.

Mr. Antonino da Silva Neves of the Republic came to India in 1920, on a mission, to study Indian agriculture and commerce. The author was earnestly requested to go over to Brazil for a term at least of two years. But owing to family considerations, he was unable to comply with the request very kindly made. However, with pleasure, we rendered all assistance to Mr. Neves. We took him across the jute districts during the growing season, 1920. Different crops of jute of different tracts of the country were all shown to him. We procured for him seeds of some typical races. He despatched over 100 maunds of the seeds thus collected to his country for sowing during the season beginning in November, which is identical with the jute season in India commencing from April.

Many friends of the author took objection to his helping Mr. Neves in this respect, as they believed that it would harm India if jute were successful in other countries. But in these days of democracy, no country should have the right to monopolise a crop of the world's commerce. This, however, is a matter of opinion.

The jute industry has already made astonishing progress in Germany, France, United States and Italy. Before the war, Germany was, next to the United Kingdom, our best customer for raw jute. Now, Germany sometimes tops the United Kingdom. The industry is being steadily introduced into Italy, Belgium, Spain, Egypt and Japan. Japan is the only country in the East which imports raw jute amounting to about 16 thousand tons annually for the manufacture of gunny bags and cloth. The first shipment of raw jute to Japan was in 1886-87 (1,447 cwts.), and to the United States about the middle of the nineteenth century.

As an experiment, jute has been cultivated in the Sudan. How far the country has been successful, in the introduction of the plant, is not known. From the results of chemical examination of a sample of jute produced there, conducted by the Imperial Institute, London, in 1911, it appears that it is inferior to Calcutta jute of finer quality, and it was pronounced to be less durable, on account of its more hydrolysis losses, although it possessed good

colour and lustre. The results of the examination are as follows :—

	Sudan Sample.	Calcutta Sample.
Moisture	9·2 per cent.	9·6 per cent.
A—Hydrolysis loss ...	10·7 ,,	9·1 ,,
B— ,, ,, ...	17·0 ,,	13·1 ,,
Acid Purification loss ...	2·2 ,,	2·0 ,,
Loss on boiling with water for 5 minutes ...	1·3 ,,	——
Ash	2·2 ,,	0·7 ,,
Cellulose	72·1 ,,	77·7 ,,

In the circumstances narrated in the preceding paragraphs, there is no immediate cause for fear, for the crop being introduced into other countries. They will, at first, have to make experiments on the crop for a number of years. Secondly, a country will have to meet its own demand if the crop be a success there; and thirdly, the question of exports will arise. Cheap production of the fibre in India is in its favour to compete successfully with it or any other fibre that may be produced elsewhere.

CHAPTER III

Forecasts of Outturn

FORECASTS of outturn of jute are published by Government twice in a season. The preliminary forecast estimating the area sown with the crop and giving a description of weather conditions of the season for sowing and also the general conditions of the standing crop, is out in the middle of July, and the final forecast is published early in September, instead of the middle of September as it was the case before. The final forecast deals with the change in the area, if any, and the estimated outturn and the weather conditions during the growing period. The estimated outturn is the principal item of the report with which the trade is vitally concerned.

The forecasts are prepared by the offices of the Provincial Directors of Agriculture on the basis of reports submitted by the District Magistrates, who collect the information through the agencies of the Police. The figures of the forecasts, as estimated lately by the Departments, appear to be far below the actuals, with exception to that of 1924-25. The short estimates may be due either to underestimated areas submitted from the districts, or to the estimated normal average outturn of the crop, at 3·3 bales, per acre. In our opinion, the yield of crop as estimated by the Departments, at 3·3 bales or thereabout at 16 maunds, per acre, is fairly correct. But the estimates of the acreage by the district officials, we believe, are underestimated.

There is an influential section of the public who are of the opinion that these incorrect forecasts are not of any use to the cultivators, if they are not exploited of their weak position by the buyers, and therefore the Government is persistently requested by this section to abolish the publication altogether. In our opinion, this prayer is not justifiable, on the grounds that in modern times, the publications of the reports showing the true state of agricultural conditions and estimated outturns of crops of a country are indispensable for its own use and also for international purposes. These reports are to help both the growers and the consumers in order to adjust the supply and the demand

of the world. The memorialists would have done well, if they had urged Government to re-organise the arrangements for publication of accurate reports and the distribution of vernacular editions among the cultivators.

The first important factor, in preparing an estimate of outturn, is the area under crop, in a year. Unfortunately there is no Government agency in a village, in the jute growing provinces, to report on the area cropped accurately. Police reports are not of much value, unless they are thoroughly checked by experienced officers of the Agricultural Departments. The area reported in the Survey and Settlement Reports cannot also be taken as accurate, as far as jute is concerned, for the area under jute varies considerably from one year to another, and secondly, as during Survey Operations the paddy stumps which followed jute are found on some fields, this area is shown under paddy. Paddy is transplanted in the same fields, just after jute is cut, in some parts of the country. However, the portion of the area under jute in relation to the whole area cropped, may be roughly estimated by experienced men, if they know the figures of the nett cropped area of a district which has been surveyed. Lately, efforts are being made to get reliable estimates of the area under jute, through the village *Panchayets*. The *Panchayets* prepare schedules showing the names of the cultivators of a village and the estimated areas sown with jute by them. Unless the fields are actually surveyed after the seed is sown, the estimate cannot be taken as correct. The author checked, on many occasions, the figures thus submitted by the *Panchayets* and found that the *Panchayets'* figures were mostly less than actuals.

The raiyats are suspicious of any work undertaken by the Government. They would hardly give accurate figures to the *Panchayets*. From financial consideration, it would not be possible for Government to employ an *amin* (surveyor) in each village for preparing crop reports. In these circumstances, until the raiyats have confidence in the officers under the Government, there is no other way that can be suggested for preparing agricultural returns. As a first step, to inspire confidence, we would suggest to re-organise the village Unions or societies, taking a representative from each family. The co-operative societies should be amalgamated with them. They will lend money to the deserving men. They will decide petty disputes both civil and criminal, as at present done, and will help Government gradually in every

way. They will also sell all the produce of the village and take, say, one anna from each rupee sold. So, after a few years, funds will be accumulating from which money can be lent to the raiyats at a nominal rate of interest. In this way, the oppression of the *mahajans* or *banias* will cease, and the petty traders called *farias*, who are responsible for fraudulent watering of jute and adulterating other crops, will be extinct. In a word, all the corruptions of the rural country will disappear, in a short course of time, and the country will be prosperous. It is hoped that both the Government and the people will combine to form such societies. These re-organised unions would be most useful institutions for supplying Government with accurate information relating to area and outturn of crops, and to all other matters in which the people and the Government are concerned.

The next factor, for estimating outturn, is the *standard normal outturn*. This normal outturn of a crop, per acre, on an average soil, in a year of an average character of seasons, is ascertained by Government from past experience, on the basis of various crop-tests, in various parts of a province, conducted by Government officers. The figures of average outturns are ordinarily revised by Government once in five years.

There is another factor for calculation of outturn. It is the *fraction representing the relation of the crop reported on to the normal crop per acre*, what is known as *percentage estimate*. The cultivators estimate the crop outturn in annas or sixteenths—sixteen annas representing the full crop. A crop above sixteen annas, according to the raiyats' estimate, is a bumper crop. But commonly, raiyats underestimate their crops. Sixteen annas are, therefore, taken to represent a bumper crop, while twelve annas a normal crop, that is, cent. per cent. In some parts of India, sixteen annas however express a normal yield. The normal outturn of jute in Bengal is taken at 3·5 bales per acre, 3·3 bales in Assam and 3 bales in Bihar and Orissa (1 bale=400 lbs.) When these three factors are determined, they are multiplied to arrive at the required estimate of outturn.

Thus taking the area under jute in Bengal in 1919 at 2,500,000 acres, the standard normal outturn at 3·5 bales per acre, and the *percentage estimate* at 80, the required estimated outturn may be worked out as follows:—

$$2{,}500{,}000 \times 3{\cdot}5 \times \tfrac{80}{100} = 7{,}000{,}000 \text{ bales.}$$

The crop forecasts are issued in English which never reach the cultivators. They are quite in the dark as to why the information regarding area, etc., under a crop is wanted by Government. This is the main reason of their being suspicious. The object and usefulness of forecasts may be made known to the cultivators if vernacular editions of the publications be distributed among them. Thus they would be able to judge for themselves that it is for the benefit of the cultivators and traders alike, that forecasts are published.

CHAPTER IV

Paper and Paper Pulp

(Utility of Jute Stalks)

ACCORDING to modern historians, the art of paper manufacture was first introduced into India, in Kashmere, from Samarkand, sometime during the thirteenth century. From Samarkand, it is said, the knowledge of paper-making reached the Arabs early in the eighth century. Sir Aurel Stein, during his excavations of the Niya in Khotan, found a letter, dated 718 A.D., written by a Jew resident in Persia, addressed to a Khotan official. Various other kinds of paper written in Khotan and China were also discovered by him, which shows that Khotan was once the seat of paper manufacture. The art of paper-making is said to have been invented in China in 105 A.D. But it is difficult to understand why paper-manufacture was not introduced into India earlier, although close relationship, religious as well as commercial, existed between the two countries from ancient times.

The Art of Paper-making

Indian antiquarians, on the other hand, are of opinion that the art of paper-making was not unknown in India, before it was invented in China. Niarcus, who accompanied Alexander the Great, in his conquest of the Punjab, found that a sort of paper-like pasted leaf made of cotton was used for writing documents in commercial transactions. However, it is true that paper was not used for writing sacred books until recently, even if it was invented in India in ancient times or introduced later. Nicolo Conte, who visited India in the early part of the 15th century, says that the inhabitants of Cambay alone used paper. For writing sacred books, the barks of *Talish*, called *Talishpatra* (Cassia Tamala ?), and birch tree (*Bhurjapatra*), and afterwards the leaves of palmyra palm (*Talpatra*) were used. There is no Sanskrit name for paper except *patra* (leaf),

even if it was at all known in ancient India. *Kagaj*, which is now used in the vernacular, is a Persian word. In modern Sanskrit, which sprang up after the decline of Buddhism sometime during the seventh century A.D., a word *Kagada*, conveying the same meaning as *patra*, came into use.

Introduction in India

Whether *Kagaj* is derived from *Kagada* or not, we cannot say. It is true that the industry was introduced throughout India by the Emperor Akbar, probably after his conquest of Kashmere, which was then a famous paper-manufacturing centre. Since then, the industry flourished in many places, Maldah, Rangpur, Dacca and Faridpur in Bengal being the most important among them. The paper was manufactured chiefly from waste paper, old rags and ropes. It was called *Tulat Kagaj* in Bengal, that is, the paper made of *Tula* (cotton). It is not known why it was so called, for cotton, as far as we could ascertain from the families who manufactured paper before, was not used. It may, however, be allowed to pass that it was manufactured formerly from cotton also. Bamboo pulp has been used for paper-making in Nepal since a long time.

Hand-made Paper

Until 1880 or so, the hand-made paper industry was in a flourishing condition in Bengal, before steam power machinery was introduced. As a result of the *Swadeshi* agitation, some of the descendants of the old paper manufacturers—a class of Mahomedans known as *Kagajia*—were induced to revive the industry. They made but little progress, due to competition with cheap and better material either imported or manufactured in India by steam power.

Mill-made Paper

The quality of paper manufactured in the Indian mills has of late greatly improved. But it is still inferior to that imported from abroad. The yielding capacity of the existing mills is similarly unsatisfactory. In 1917 there were only ten mills in India, of which four were in Bombay, three in Bengal, one in the United Provinces, one in Travancore and one in Gwalior. The authorised capital was estimated by the authorities to be at about Rs. 53,00,000 only. The production of paper of these mills was estimated at 31,900 tons in 1917, which was much less than half of the Indian

demand, the total demand being about 84,000 tons. During the year 1931-32, there were three mills in Bengal, three smaller mills in Bombay and one small mill in the United Provinces, in India, and the total import of paper into India was estimated at 41,040 tons, 75 per cent. being printing paper.

Pulp Problem

The question of expansion of paper manufacture in India is receiving serious attention from the authorities, and various suggestions are offered by scientists to solve the problem. The supplies of cheap raw material for paper-pulp are getting short everywhere. The woods from which paper-pulp is manufactured in the United States are also getting dearer day by day. India is fortunately rich in raw material. Bamboo alone can meet the world's demand for a number of years, and it is abundantly available in the forests of Assam and Burma.

Raw Materials

The cost of procuring bamboos and other raw materials that can be utilised for the manufacture of paper-pulp should now receive our consideration. The cost of cutting bamboos in the forest, and that of transporting to a mill, may be estimated at about Rs. 20 per ton, the rate at which it is being sold in the towns for building and other purposes. *Nal* (elephant grass), which is also suggested for manufacturing paper-pulp, may be had at the same price. Although it grows wild in the swamps of Bengal, it can supply only a limited portion of the expected demand. *Munj* (*Saccharum arundinaceum*) would be more costly as it is not available in large quantity. The *Baboi* (*bhabar, sabay*) grass, which is at present used in the Indian mills for paper-pulp and which was very cheap some time ago, is now sold at Re. 1-5-0 per maund. The quantity available is six to seven lakhs of maunds or over one lakh of bales. Paddy straw is now available at about Rs. 15 per ton. Paddy straw contains a considerable quantity of silicates which make it very difficult to reduce to paper-pulp. This stuff is utilised for paper-making in Germany only. In the United States, on the other hand, special powerful machinery has been introduced to get paper-pulp out of timber. Before the introduction of special machinery for treatment of the cuttings and rejections of jute for manufacture of gunny bags, these materials were largely

used for paper-making, as they were very cheap then. The other materials, such as old rags and waste paper, which were formerly obtained at cheap prices, are not now available easily.

Value of Jute Stalk

We may suggest the name of a material by way of making experiments. It is the stalk of jute, a bye-product of the jute fibre which is now either wasted or burnt as fuel. It is not valued much as fuel as it is very soft and burns quickly. In point of woody fibre it compares very favourably with other raw materials used in the manufacture of paper. Nay, it is much richer than any other materials mentioned above and it may be had at a nominal cost.

Chemical Analysis

The results of chemical examination of the jute stalk and some other raw materials conducted by us are tabulated below :—

Name of Article.	Fibre.
Jute Stalk	60·50 p.c.
Bamboo (ordinarily available in Calcutta)*	41·00 p.c.
Sabay Grass	38·00 p.c.
Nal	37·00 p.c.
Paddy Straw†	33·00 p.c.

What Jute Stalk Contains

The results of complete analysis of the jute stalk will, it is hoped, be of much interest. They are as follows :—

Water (Hygroscopic)	...	5·85 p.c.
Protein ...	...	1·40 p.c.
Ash	...	0·92 p.c.
Carbo-Hydrate ...	...	31·33 p.c.
Fibre	...	60·50 p.c.

* Containing 3·20 per cent. of ash.

† Containing about 8 per cent. of ash.

Its Vast Possibilities

The fibres of the above-mentioned materials were estimated by sulphuric acid and sodium hydrate treatments. The jute stalk, it will appear, contains only a negligible quantity of ash, which is easy to separate from the fibre (pulp). It contains little colouring matter, requiring therefore little bleaching operation. About a ton and a half of the jute stalk, per acre, would be available as a bye-product of jute fibre cultivated in this country. Thus the jute districts of Bengal, Bihar and Assam would yield annually about 40 lakhs of tons of this raw material, producing roughly about 20 lakhs of tons of paper-pulp. If jute stalk be successful as a paper material, it will, it is hoped, like its main product, revolutionise Indian industries and commerce.

Hand-made Paper of Bengal

The art of hand-made paper manufacture, which was once in a flourishing state in Bengal, is now almost forgotten. From our enquiries on the subject during 1920, it appeared that there was only one family out of about 1,000 at the village Arial in Dacca, who still manufactured paper. Paper was also manufactured by a few families near Tarakeswar in Hoogly. They used waste paper, rags, old ropes, gunnies, gunny cloths, nets, etc., as paper materials. These were cut into small pieces and put into earthen vessels containing water. After a week or so, the stuff was washed in a basket in order to get rid of the dirt contained. Then it was again immersed in water strongly impregnated with equal quantities of *Saji-mati* (crude sodium carbonate) and quicklime. After a week or ten days, it was again washed, broken by beating, and dried. This process was repeated three to six or seven times, before the pulp became fit for making paper. Pulp is easily made out of waste papers. Rice gruel is used for sizing.

Hand-made Paper of Europe

In European countries good hand-made paper is still manufactured out of rags. The rags are at first dusted and cut into pieces, and then washed to get them free from dirty matter. The stuff is then boiled in caustic soda solution of 5 per cent. to 10 per cent. in strength. Too weak lye leaves behind a part of substances to be eliminated : if too strong it injures the tenacity of the fibres. It should,

therefore, be neither too strong, nor too weak. Ordinarily soda and lime process is resorted to as it is cheaper. For 100 parts of rags about 1 part of soda and one-half to 2 parts of unslaked lime are used for preparing lye in 100 parts of water. Half the quantity of the solution is generally used in the beginning of boiling and the other half after six hours of boiling. If necessary this process is repeated for the second time. Slaked lime is dissolved by sprinkling with a little water, then mixed with soda in the preparation of the lye.

Beating

After boiling, the stuff is washed in clean water and bleached if necessary. Next comes beating for about six to eight hours to reduce the stuff to fineness, and pressed to remove excess water.

Sizing

The stuff is now to be sized. There are various ways of sizing, animal or vegetable. The animal size is gelatine or glue, and the vegetable size is made of a combination of resin with soda. Alum is used to assist in fixing the size in the paper.

Filling

Mineral matters are employed for filling or loading. China clay is the principal, others being Barium-sulphate, Calcium sulphate, Magnesium silicate.

Colouring

Colouring matters are required for the majority of papers. Small quantities of blue and red colours are used for white papers, and aniline dyes as well as various pigments for coloured papers. Finally the stuff is made into paper.

Paper Manufacture by Steam Power

In the same process paper is manufactured by steam power by which the work is done far more quickly at much less cost. In the mills vacuum pans are employed for boiling.

Fibres for Paper-pulp

Rags are the best as a paper material. But, for cheapness, all kinds of vegetable fibres are utilised for paper-making. Cellulose is the substance required for paper. It is found more or less in all tissues of plants. In most cases the presence of colouring matters and various waxy and resinous substances render the cellulose very impure. It is, therefore, desirable that all impurities be removed, as far as possible, before the fibres are made into paper. Tenderest rags, such as muslins, are employed for blotting papers. Jute fibre is smooth but difficult to bleach: however, the resulting paper is strong and tough.

The length of vegetable fibres, varying from $\frac{1}{25}$th to $\frac{1}{8}$th inch, operates an important part in the manufacture of paper. Short length of a fibre weakens the quality of paper. Papers made out of straws are weak. Straw pulp is generally used as an admixture. Wood pulp is now in common use, specially in America.

Straw and wood are reduced to pulp by steam power by either Soda-lime or Caustic soda process. Caustic soda is preferred. Wood is to be cut into chips after removing its bark and grasses into pieces before boiling. These stuffs may also be reduced to pulp by boiling over a naked fire in stationary cauldrons unlike rotatory boilers used in the mills. But it will take much time and labour before they will be reduced to pulp.

We tried to induce some of the mill owners to give a trial to the jute stalks, but they were not prepared to do it as it would then require new machineries and re-adjustment of things.

CHAPTER V

Three-hundred-Acre Jute Farm

WE need hardly repeat that jute will never succeed if sown under unfavourable conditions. We have accentuated these points on several occasions, in the preceding pages. Jute grows on particular soils, under particular climatic conditions. Sowing and after-treatments must be finished in proper time. Delay in these treatments will harm the crop materially, and the grower will sustain a loss. Although there is a sudden fall of prices, yet a well-grown crop should pay its grower, as wages for labour have been similarly reduced to between annas 4 and 5, per head, per diem.

Of the 300 acres of a farm, 150 acres may be sown with jute and the other 150 acres with paddy, in the same season. Jute may be rotated with rape seed or pulses or tobacco in the first year, while paddy with jute in the second year. An additional area of 50 acres is also wanted for the farm for buildings and grazing.

The following statement showing receipts and expenditure of a three-hundred-acre jute farm, is estimated, according to present circumstances.

RECEIPTS :—		Rs.
3,000	Maunds of jute at Rs. 5 per maund (20 maunds per acre)	15,000
3,000	Maunds of paddy at Rs. 2 per maund	6,000
1,500	,, of mustard at Rs. 4 per maund	6,000
7,500	,, of *pekati* (stalks) at about 2 annas per maund (50 maunds per acre)	1,000
3,000	,, of paddy straw at 3 maunds per rupee	1,000
4,000	,, Juar plants for fodder ...	500
		Rs. 29,500

The value of the outturn as has been at present estimated is reduced nearly to half of the former value.

Capital Expenditure :—		Rs.
Cleaning and plotting		6,000
Purchase of 120 bullocks		6,000
" implements		2,000
Cattle sheds, quarters, etc.		6,000
	Rs.	20,000
Working capital	"	5,000
	Total Rs.	25,000

Recurring Expenditure :—		Rs.
60 Ploughmen at Rs. 10 each per mensem ...		7,200
6 Sirdar coolies at Rs. 10 each per mensem		720
6 Servants for feeding and grazing cattle at Rs. 10 per head per mensem ...		720
Extra labour charges		7,000
550 Maunds of oil cake for bullocks at 1½ maunds per diem		1,100
36 Maunds of salt for bullocks ...		100
2,000 " of paddy straw		700
4,000 " of juar plants		500
Seeds (jute, paddy, mustard, etc.)		1,000
Depreciation on bullocks and implements ...		600
Rent at Rs. 3 per acre		1,050
Miscellaneous		810
	Rs.	21,500
Establishment charges	Rs.	3,000
Total Recurring Expenditure ...	Rs.	24,500
Total Receipts	"	29,500
Net profit (on a Capital of Rs. 25,000) ...		5,000

CHAPTER VI

Kerosene Emulsion

KEROSENE emulsion is the most simple and effective remedy against all insect pests. It kills insects when applied to their bodies. In it's raw state, Kerosene oil injures the plants, so it should always be applied after being diluted with some other material which would form an emulsion in mixture with the oil. It can be used by means of an ordinary hand sprayer in small plots or gardens. For a large area under field or garden crops, a larger machine of Knapsack Sprayer pattern would be very useful. It is especially valuable when delicate plants are attended to. The sprayer which is used commonly by the Municipalities for disinfection will do well for ordinary purposes. Emulsion may be prepared according to the following formulæ :—

Soap Formula

Kerosene oil	2 gallons.
Fish-oil-soap	$\frac{1}{2}$ lb.
Water	1 gallon.

The soap, cut into small pieces, is to be dissolved in the water by boiling and immediately added, away from the fire, to the kerosene oil. The whole mixture is then agitated violently while hot with an ordinary butter churn used in this country. The emulsion should be ready after 5 or 6 minutes' agitation, when it assumes the consistency of cream. The emulsion should be diluted with water when it is to be used. If fish-oil-soap or soft-soap (potash soap) be not available, ordinary bar soap may be used.

Up-country, where the water is hard, some of the soap will combine with the lime or magnesia in the water, and the oil will be freed more or less, especially when the emulsion is diluted. Such water should be broken with lye, or rain water employed. It would be better to follow the " milk emulsion " formula, with which the character of the water, whether hard or soft, does not affect the result.

Milk Formula

Kerosene oil ... ... 2 gallons.
Sour milk ... ... 1 gallon.

The oil and sour milk are to be mixed together and churned as in the former case. After about 5 minutes' agitation, a change from a watery liquid to a thick buttery consistency takes place, which is thicker than that with the soap. Sweet milk may also be used for the sour milk after a little vinegar is added to it. Milk emulsion cannot be stored for a long time. It would be better to prepare it when required.

The Emulsion is to be diluted

For the soft-bodied insects and plant lice, one part of the emulsion is to be diluted with from 10 to 15 parts of water ; for the spider and other plant mites, the same, with an addition of 1 ounce of flowers of sulphur to a gallon. For scale insects, the larger plant bugs, larvæ and beetles, it should be diluted with from 7 to 9 parts of water.

CHAPTER VII

Land Measures and Bazar Weights

IN several places of this book, local measures of lands and standard *bazar* weights have been quoted. A purchaser should know local weights, which vary at different markets, before prices for commodities are settled. The diversity is still more confusing in case of land measures in this country. Almost every Zemindari (an estate) has its own standard measurement. The unit of measures of lands is a *nal* (reed) in Bengal or *laga* (a piece of bamboo) in Bihar, which may vary in length from 4 cubits to 8 cubits —18 inches making a standard cubit now-a-days. The *bigha* measurement is in vogue in Western and Central Bengal. Eastern Bengal has both *bigha* and *kanee* measures, while in Northern Bengal it is known by the name of *done*.

The principal common measures are given below :—

Land measures

English—

12 inches=1 foot
3 feet=1 yard
$5\frac{1}{2}$ yards=1 pole
22 yards (4 poles)=1 Gunters chain
1 square chain or 484 square yards } $=\frac{1}{10}$th acre
4,840 square yards=1 acre
640 acres=1 square mile.

Bengal—Standard bigha

1 square cubit=1 ganda
20 gandas or 5 cubits × 4 cubits } =1 chatak
16 chataks=1 cottah
20 cottahs=1 bigha
or
4 cubits=1 nal
1 nal × 20 nals=1 cottah

Bengal—Standard bigha—*contd.*

20 nals × 20 nals or
1,600 square yards } = 1 bigha

3·02 bighas or
60$\frac{1}{2}$ cottahs } = 1 acre.

1$\frac{3}{5}$ Bihar bighas
(5$\frac{1}{2}$ cubits making a *laga*) } = 1 acre

Done = 13 cottahs or $\frac{1}{5}$th acre
Kanee = 3 bighas and 12 cottahs or 1·16 acres
16 Kanees = 1 drone

In the district of Tippera 1 drone is equal to 6$\frac{2}{5}$ acres or about 19 bighas.

Kiari or Kiar = nearly 1 bigha
Pakhi = 0·8 bigha
Hal = 15 bighas or 5 acres

Bazar Weights

English—

16 drams = 1 ounce (oz.)
16 ounces = 1 pound (lb.)
14 pounds = 1 stone

2 stones or
28 lbs. } = 1 quarter

4 quarters or
112 lbs. } = 1 hundredweight (cwt.)

20 hundredweights or
2,240 lbs. } = 1 ton

Bengal—Standard weight

5 tolas = 1 chatak (2$\frac{1}{4}$ oz.)
4 chataks = 1 powa
4 powas = 1 seer (sr.) [2$\frac{2}{35}$ lbs.]

In some places 60 tolas and somewhere 90 tolas make a seer.

5 seers = 1 panseri
40 seers = 1 maund (md.) [82$\frac{2}{7}$ lbs.]

27·22 maunds or
27 mds. and 9 srs. } = 1 ton

CHAPTER VIII

Bengalee Year

BENGALEE year ends with *Chaitra,* corresponding to the months from the middle of March to the middle of April, after harvesting and storing the crops of the year. New year begins in *Baisak*—the most auspicious month of the year. Social as well as religious festivals are held almost daily, during this month, in all parts of the Province—miseries and failures of the past year being forgotten. Old traders renew their books and new-comers step in. After a year's hard work, the jolly peasant of Bengal enjoys heartily the merriments of the month for a few days and commences field-work over again with greater energy. His crops are only a few and the area is but limited. Hence, every plot under cultivation receives his careful attention, although it is carried on in a primitive style. Jute is a special favourite of his, as it secures a loan for him readily, when he is urgently in need of money. The present slump in the jute trade has no doubt made the situation dull, he is however waiting for better days. Jute as well as other substitutes are being vigorously tried in other parts of the world, but up to now they have been unsuccessful. We wish good luck to the growers, the traders, and the manufacturers—all who are concerned, in any way, in the crop about which this book is written.

Bengalee months were frequently quoted in this work for convenience sake, when information regarding the seasons related to sowing or harvesting was collected from the cultivators. The season for trade in jute begins in July and ends in June. The table following gives the Bengalee months corresponding to the English months.

Bengalee month		Corresponding to English.		
Baisak	from middle	April	to middle	May.
Jaistha	,,	May	,,	June.
Ashar	,,	June	,,	July.
Sravan	,,	July	,,	August.
Bhadra	,,	August	,,	September.
Aswin	,,	September	,,	October.
Kartik	,,	October	,,	November.
Agrahayan	,,	November	,,	December.
Poush	,,	December	,,	January.
Magh	,,	January	,,	February.
Falgoon	,,	February	,,	March.
Chaitra	,,	March	,,	April.

CHAPTER IX

Glossary

A*CTUALS*—Specified marks of jute, guaranteed to be of some standard qualities, shipped during the two previous seasons, and recognised by the Calcutta Baled Jute Association, the London and the Dundee Jute Associations.

Aratdar—An Indian commission agent who sells goods sent to him by *mahajans* or *beparies*, at his own discretion, if no other special agents are appointed by them to look after their sales. The *aratdar* charges a commission generally at the rate of one anna and three pies for every maund of jute sold, besides actual godown rent.

The aratdar is always responsible for the recovery of the money for goods sold. He now and then gives advances to his respectable *beparies*, for which an interest of 12 per cent. per annum is charged. An *aratdar* must be a rich person.

Ash mara—Weak fibre ; literally *ash* means fibre and *mara* dead.

Ash mota—Coarse fibre ; *mota*—coarse.

Baler—A jute dealer who buys loose jute and, after assortments, makes it up into bales of 400 lbs. each of the quality or qualities thus assorted, separately, and sells his stock either to shippers or the local mills for manufacture of gunnies or gunny cloth.

Barky—The fibre containing epidermic bark which is not removed in washing is called barky.

Bach pat—The fibre from immature plants rejected at the time of thinning.

Bepari—An Indian trader who has no godown of his own. He carries on his business at the places of *mahajans* or *aratdars* for which he pays a commission at the rate of about 2 annas per maund.

In the Calcutta Bazar, the *bepari* pays the buyers ½ anna per maund as brokerage, and Re. 1 per Rs. 1,000 as cashier's fee, and 4 annas for every one hundred drums as *jalpani* (tiffin charge).

Bheel—A small lake commonly overgrown with weeds and grasses containing abundance of fish. In course of time, it is filled with earth and rotten vegetable matters, to be fit for cultivation.

Bombays—A group of baled jute contained of hard, short and barky fibre with or without tap roots.

Broker—An agent employed to effect bargains or contracts between buyers and sellers for which he gets from the latter a commission called brokerage at the rate of 1 per cent. of the value of goods sold. In the Calcutta Bazar, he gets 3 to 6 pies per maund of the loose jute sold.

Buk-chhal—Fibre containing bark (*chhal*) due to the plants being grown in water for some days, causing a defect for which the bark is not fully removed by retting and washing.

Burlap—American name of the *hessian* cloth.

Butts—The short hard root ends of jute cut off.

Camb—An apparatus comprising the heddles which shift the sets of warp threads alternately.

Champadar—One who places jute of specified quality over the scales for weighment.

Char land—The land on the banks of rivers.

Chat—Coarse hessian cloth.

Croppy—Fibre having top ends rough and hard.

Cuttings—*Cuttings* are the portions of jute cut off from the root or top ends of fibre which is hard, mossy or barky at the bottom end, while weak at the top end.

Dagie—Damaged.

Dalal—A Bengali term for a broker.

Dana—The quality of the jute fibre in regard to fineness or coarseness or its strength. Literally *dana* means a grain.

Dhaltha—Loss of fibre on drying, on account of excess moisture contained by it. The *bepari* has to offer one-half to one seer of the fibre per maund to the buyers as *dhaltha* to make up the loss. There are markets where the jute is free from moisture and no *dhaltha* is charged.

Diarah land—Same as char land.

Dollop—A fixed weight of raw jute to be spread in a given time, on a Breaker Feed Cloth.

Double warp—Two threads going into one heddle for weaving.

Faria—A petty Indian buyer who goes about from house to house of the raiyats or to the interior markets for buying jute to be supplied to the higher buyers.

Feed-arpon—Canvas cloth, same as Feed cloth, adjusted on a frame in front of a Breaker for feeding fibre to the machine.

Fill—Weft thread.

Fine—Fibre of superior quality, of which the *dana* is not coarse or thick, and being free from impurities.

Flabby—Wanting in firmness in the stick of the fibre, that is, loose.

Fulpat—Immature stuff cut when it begins to flower. This fibre is excellent in colour, but weak at the top.

Ghatta—The fibre which is tangled by too much handling so as not to be easily separated.

Glossy—Denoting the quality of a class of jute which possesses a good lustre.

Gun (Goon)—Same as *chat* of finer quality known as *gun* from which the term gunny is derived. Originally *gun* was a name for hemp thread or string mostly used for weaving a mattress.

Gunny—The present general term for hessian or sacking cloth or for gunny bags.

Habijabi—Stray fibre of the assortment room jumbled together.

Harness—Same as *Camb.* An apparatus comprising the heddles through which the warp threads pass, as one harness lifts another drops, when a shuttle is shot across between the alternate threads.

Heart-damaged—Rotten fibre due to fraudulent watering inside a drum or a package and stocked without being noticed by the buyers in due time for drying.

Heavy goods—Coarse cloth or sacks.

Hessian—Gunny cloth of finer quality.

Jachandar—A person who examines the quality of the fibre of every bundle to be weighed, on behalf of the buyers.

Jhump—The Bengali term for *Camb.* Literally it means a jump.

Jews mallow—The species of *C. olitorius* (Deshi pat) was originally known in Europe by this name.

Kapali—A Bengali sack-weaver by caste.

Knotty—Full of knots. Knot is a portion of fibre agglutinated which resists separation, mainly due to an insect bite or puncture on the growing plant.

Koyal—An employee of the *aratdar* who makes weighment.

Kutcha Bale—A loosely pressed bale intended for despatch to the Indian Mills or to the Calcutta Bazar.

Mahajan—An Indian trader who buys jute either from raiyats or *farias*. He generally sends his goods in drums to an *aratdar* in Calcutta for disposal. He must have godowns of his own. If he has any *bepari*, he must help him by all means. A *mahajan* is also called *bepari* by the *aratdars* if he sells his goods through them.

Mora-head—It is a term applied to a rejected class of straight jute (against *ghata*) not too short to take a turn at the head after a twist (*mora*).

Mossy—A low land jute which had numerous adventitious roots sprung up when the plants grow in swamps.

Ojan Sarkar—A clerk who notes the weights of bundles when weighment is made.

Paikar—Same as *Faria*.

Porter—A term used in the mill weaving denoting number of warp threads per inch of a woven fabric, and that of the weft is called a *shot*.

Pucca Bale—A strongly pressed bale by steam power containing 400 lbs. of fibre intended for export. Each bale is to measure 10 c.ft. in volume.

Raiyat—A petty Indian farmer.

Retting—A process of decomposition of epidermic bark of the plant immersed in water for separating the fibre from the woody stump.

Rooty—Jute containing roots.

Sacking—Coarse woven jute cloth intended chiefly for gunny bags.

Shot—*Vide porter*.

Shuttle—A weaving implement about 9″ long and 3″ broad, pointed at both ends, by which weft thread is shot across between the warp threads.

Skein—A bundle of yarn or thread made by coiling. A skein of cotton yarn is 360 feet in length, formed by eighty turns of the thread round a 4′-6″ reel.

Snipping—Same as cutting the root and top ends of the jute fibre.

Specky—Having patches of outer bark on the fibre.

Spinning—The act of drawing out and twisting into thread.

Sticky—Fibre containing broken pieces of sticks (stalks).

Substitutes—Marks of the grade and quality equal to those of *actuals*, of which they are substitutes.

Tant—A common Bengalee term for a loom, derived from the Sanskrit word *tantu* (thread).

Twill or *tweel*—A textile fabric, woven with 3 *cambs*, in which weft threads pass alternately over one warp thread and under two or more, producing diagonal lines.

Twist—Two or more threads twisted together.

Under-broker—A broker who is not allowed to approach the buyers, but effects business through a broker. The broker allows him a share of 25 per cent. in the brokerage.

Warp—Threads stretched lengthwise in a loom to be crossed by *weft* threads.

Weaving—The art of manufacturing cloth, or any other fabric, by interlacing threads.

Weft / *Woof*—Weft or woof is a thread that crosses the warp threads in weaving.

Yarn—Spun thread.

STATEMENT I.

Area and Yield of Jute Crop according to Government Forecasts and Actual Exports and Purchase by Calcutta Mills. Figures in Thousands, during season, July to June.

YEAR.	According to Government Forecasts.		*Exports of Raw Jute from Bengal, Bales.	Mills' Purchase, Bales.	Mills' Actual consumption, Bales.	Gross Total (including 500 thousand bales estimated to be the country's consumption).
	Area, Acres.	Yield, Bales.				
1931-32	18,99	58,45	27,80	......	......	......
1930-31	18,61	55,66	34,27	62,26	44,37	101,53
1929-30	34,92	112,55	44,46	59,62	62,46	109,08
1928-29	34,15	103,85	44,28	55,55	58,79	104,83
1927-28	31,44	99,16	44,86	61,64	56,33	111,5[illegible]
1926-27	29,62	121,87	44,48	74,35	53,74	123,83
1925-26	29,26	90,00	35,16	53,35	53,44	93,51
1924-25	27,70	81,20	38,22	47,89	55,19	91,11
1923-24	27,87	84,73	37,71	51,16	50,04	93,87
1922-23	18,00	54,86	29,02	29,87	46,15	63,89
1921-22	15,18	40,65	29,68	44,54	42,36	79,22
1920-21	25,09	59,79	23,43	50,45	54,67	78,88
1919-20	28,39	85,41	33,58	60,89	50,82	99,47
1918-19	25,00	70,19	21,93	49,43	49,97	76,36
1917-18	27,36	89,29	17,26	61,42	52,96	83,68
1916-17	27,03	88,80	28,10	53,50	55,20	86,60
1915-16	23,76	74,24	31,06	53,28	56,09	89,34
1914-15	33,59	110,53	29,67	60,10	48,05	94,77
1913-14	31,70	[illegible]8,75	41,92	43,95	43,74	90,87
1912-13	29,70	[illegible]9,84	49,42	47,41	44,35	101,83
1911-12	31,06	[illegible]8,32	46,26	41,62	37,56	92,88
1910-11	29,38	79,32	35,31	36,72	39,80	77,03

* Of the exports of raw jute, roughly about 25% being *Cuttings* and 25% Rejections.

STATEMENT II.

Area and Yield of Jute (Government Forecasts).

Figures in Thousands.

YEAR.	BENGAL (including Cooch Behar and Tripura States).		BIHAR and ORISSA.		ASSAM.	
	Area under Jute in acres.	Yield in Bales of 400 lbs. each.	Area under Jute in acres.	Yield in Bales of 400 lbs. each.	Area under Jute in acres.	Yield in Bales of 400 lbs. each.
1932 ...	16,33	51,28	1,57	4,33	1,09	2,84
1931 ...	16,14*	50,03	1,48·8	3,67	99·3*	1,97
1930 ...	30,52	99,66	2,38	6,70	1,92*	6,19
1929 ...	29,89	91,90	2,38	7,19	1,57	3,52
1928 ...	26,70	85,19	2,47	6,93	1,95	6,24
1927 ...	29,33	90,04	2,41	6,67	1,71	4,66
1926 ...	33,21	106,52	2,97	7,64	1,86	5,99
1925 ...	26,85	79,51	2,63	6,40	1,36	2,79
1924 ...	23,58	71,66	2,46	4,89	1,34	3,33
1923 ...	24,10	74,63	2,23	5,28	1,20	3,39
1922 ...	15,28	47,46	1,60	3,43	90	2,67
1921 ...	13,16	35,95	1,08	2,25	81	1,54
1920 ...	21,69	52,47	1,79	3,34	1,18	2,59
1919 ...	24,59	75,68	2,03	5,12	1,37	3,32
1914 to 1918 (Great War Period, average)	23,81	74,89	2,23	5,72	95	2,40
1911 to 1913 (Pre-war Period, average)	25,77	79,62	2,91	7,00	98	2,32

*Revised by the Provincial Departments.

Estimated Normal Outturn of Raw Jute, per acre, for :—

Eastern Bengal	... 3·7 bales.	Average Bengal	... 3·5 bales.
North Bengal	... 3·5 ,,	Average B. & O.	... 3·5 ,,
West Bengal	... 3·2 ,,	Average Assam	... 3·5 ,,

STATEMENT II—(*continued*).

Area and Yield of Jute in Indian States.

Figures in Thousands.

YEAR.	COOCH BEHAR STATE.		TRIPURA STATE.		Imported from NEPAL.
	Area, Acres.	Yield, Bales.	Area, Acres.	Yield, Bales.	Bales.
1932	21	37	1	2	...
1931	16	17	1	3	...
1930	31	80	3	8	...
1929	30	74	3	7	64
1928	32	70	...	...	72
1927	29	51	...	...	82
1926	43	1,17	...	...	40
1925	31	70	...	...	41
1924	32	74	...	...	70
1923	35	71	...	...	58
1922	22	52	...	...	57
1921	13	11	...	...	75
1920	36	58	...	...	...

STATEMENT III.

Classification of Area of Jute Districts, with reference to that of Jute.

District.	Net Area.	Net Area actually cropped.	Area under Jute.		Percentage of cultivated area sown with jute.	
			1930.	1931.		
	Acres.	Acres.	Acres.	Acres.	1930	1931
BENGAL (Out of 27 districts, jute is grown in 24 districts).						
1. 24-Parganas ...	3,107,840	730,800	60,000	45,000	9·44	6·15
2. Nadia ...	1,787,457	512,100	56,000	30,000	10·9	5·8
3. Murshidabad ...	1,369,470	623,800	33,000	14,000	5·3	2·2
4. Jessore ...	1,870,720	704,700	103,000	46,000	14·6	6·5
5. Khulna ...	3,008,491	871,900	35,000	21,000	4·0	2·4
6. Burdwan ..	1,700,329	719,300	3,000	2,000	0·4	0·3
7. Midnapore ...	3,235,200	1,806,700	8,000	4,000	0·4	0·2
8. Hoogly ...	770,320	250,300	29,000	20,000	11·5	7·9
9. Howrah ...	339,200	107,600	7,000	4,000	6·5	3·6
10. Rajshahi ...	1,676,800	697,100	85,000	68,000	12·1	9·7
11. Dinajpur ...	2,525,440	1,090,600	73,000	41,000	6·6	3·7
12. Jalpaiguri ...	1,875,840	758,000	42,000	26,000	5·5	2·7
13. Darjeeling ...	744,960	157,600	4,000	2,000	2·5	1·2
14. Rangpur ...	2,237,440	1,549,600	320,000	180,000	20·6	11·6
15. Bogra ...	882,783	443,600	103,000	60,000	23·2	13·5
16. Pabna ...	1,115,866	882,000	155,000	45,400	17·5	5·1
17. Malda ...	1,179,456	646,000	38,000	16,000	5·8	2·4
18. Dacca ...	1,759,416	1,401,800	378,000	215,000	26·9	15·3
19. Mymensingh ...	3,992,320	2,361,600	749,000	415,000	31·7	17·5
20. Faridpur ...	1,517,440	1,199,300	314,000	144,000	26·1	12·1
21. Backerganj ...	2,233,600	1,781,700	52,000	30,000	2·9	1·6
22. Chittagong ...	1,548,080	638,700	300	300	0·04	0·04
23. Tippera ...	1,648,601	1,111,300	308,000	140,000	27·7	12·5
24. Noakhali ...	969,600	740,200	64,000	28,000	8·6	3·7
Total Bengal (27 districts) ...	49,186,909	23,370,100	3,028,300	1,596,700	12·9	6·8
Cooch Behar ...	3,477,760	543,500	31,000	16,000		
Tripura State ...		......	3,000	1,000		
GRAND TOTAL ...	......	......	3,062,300	1,613,700		

STATEMENT III—(*continued*).

Classification of Area of Jute Districts, with reference to that of Jute.

District.	Net Area.	Net Cropped Area.	Area under Jute.		Percentage of cultivated area sown with jute.	
			1930.	1931.		
	Acres.	Acres.	Acres.	Acres.	1930	1931
BIHAR AND ORISSA (Out of 22 districts, jute is grown in the following 8 districts):—						
1. Champaran ...	2,263,194	1,430,200	7,680	2,500	0·5	0·2
2. Muzaffarpur ...	1,934,630	1,627,000	4,000	4,000	0·2	0·2
3. Darbhanga ...	2,142,720	1,677,800	1,236	312	0·07	0·02
4. Bhagalpur ...	2,661,120	1,684,100	4,300	2,075	0·3	0·1
5. Purnea ...	3,196,160	1,694,400	140,000	120,000	8·2	7·1
6. Santhal Parganas...	3,500,800	1,717,900	750	500	0·04	0·03
7. Cuttack ...	2,338,560	1,223,500	18,200	15,700	1·5	1·3
8. Balasore ...	1,334,400	932,200	2,800	2,400	0·3	0·2
PROVINCIAL TOTAL ... (22 districts)	53,173,099	24,958,600	179,000 (a)	147,500 (a)	0·7	0·6
ASSAM (Out of 15 districts, jute is grown in the following 9 districts) :—						
1. Cachar ...	2,467,200	365,374	200	200	0·05	0·05
2. Sylhet ...	3,497,600	1,567,050	19,800	5,000	1·2	0·3
3. Goalpara ...	2,533,760	477,596	63,300	35,300	13·0	7·4
4. Kamrup ...	2,468,170	837,108	27,100	15,000	3·2	1·8
5. Darrang ...	1,808,440	439,595	18,600	13,700	4·2	3·1
6. Nowgong ...	2,492,800	344,939	56,600	20,400	16·4	5·9
7. Sibsagar ...	331,328	730,009	400	300	0·05	0·04
8. Lakhipur ...	2,768,640	449,001	1,100	900	0·2	0·2
9. Garo Hills ...	2,019,200	98,666	4,900	4,800	5·0	4·9
PROVINCIAL TOTAL ... (15 districts)	35,299,970	5,577,979	192,000	95,600	3·4	1·7

(*a*) The Director of Agriculture, Bihar and Orissa, considers that the estimated acreage of the Preliminary Forecast, 1931, *viz.*, 148,800 acres, and that of the previous year, 238,000 acres, were correct.

STATEMENT IV.

Exports of Jute and Jute Manufactures.

Year. April to March.	Raw Jute, Tons.	Bags, No. in Thousand.	Cloth, Yards in Thousand.		Twist and Yarn, lbs. in Thousand
			Hessian.	Sacking.	
1931-32 ...	586,618	389,108	995,154	25,339	8,669
1930-31 ...	619,705	434,046	1,238,619	32,381	7,341
1929-30 ...	806,884	522,291	1,598,844	51,682	6,009
1928-29 ...	897,863	497,685	1,503,323	64,971	4,801
1927-28 ...	891,907	463,140	1,477,783	74,968	6,318
1926-27 ...	707,782	449,090	1,422,547	80,581	3,078
1925-26 ...	647,154	425,083	1,400,017	61,364	1,724
1924-25 ...	696,000	425,141	1,394,422	61,866	1,261
1923-24 ...	659,963	413,716	1,294,591	54,151	5,345
1922-23 ...	577,955	344,232	1,211,862	42,490	3,507
1921-22 ...	467,685	386,711	1,096,562	24,006	1,876
1920-21 ...	472,414	533,908	1,352,739		10,179
1919-20 ...	680,000	342,729	1,275,125		3,606
	War	Period,	average—		
1914-15 to 1918-19	512,400	665,659	1,156,113		2,680
	Pre-war	Period—			
1913-14 ...	838,500	368,759	1,061,151		155
1910-11 ...	706,200	360,880	940,101		265
1900-01 ...	708,400	202,908	365,214		2,224

STATEMENT V.

Exports of Raw Jute and Manufactures to Foreign Countries, during 1929-30.

Countries.	Raw Jute. Tons.	Twist and Yarn in thousand lbs.	Gunny Bags (a) in thousand Nos.	Cloth in thousand Yards.
Aden	......	......	330	......
Africa (East) ...	......	257	......	......
Africa (East, Portuguese)	......	......	13,552	1,391
Africa (West, French) ...	......	......	1,568	231
Africa (West, Portuguese)	......	......	13,552	......
Argentine	5,851	4,615	2,205	329,287
Belgium	46,298	34	12,491	......
Bolivia	......	......	361	......
Borneo	......	......	932	......
Brazil	18,314	......	......	162
Bulgaria	......	......	204	......
Canada	......	......	832	81,564
Cape Colony ...	......	38	5,641	......
Cape of Good Hope ...	......	......	......	919
Celebes	......	......	2,648	179
Ceylon	......	......	1,788	3,267
Chili	......	67	33,620	2,410
China	6,038	65	4,280	8,914
Colombia	......	......	193	......
Costa Rica ...	......	......	175	......
Cyprus	......	......	777	......
Denmark	......	......	1,650	......
Egypt	15,742	19,029	5,634	......
Equador	......	......	1,729	......
Fiji Island ...	......	418	......	......
Formosa	......	3,992	......	......
France	106,473	2,074	......	......
Germany	216,467	2,596	1,240	......
Gold Coast ...	......	955	......	......
Greece	1,909	3,005	1,701	......
Guatemala	......	110	......	......
Guinea	......	227	......	......
Guinea (British) ...	......	1,333	......	......
Hayte	......	2,493	......	......
Hong Kong ...	......	39,016	1,453	......
Howali	......	17,972	......	......

(*a*) One gunny bag represents 2 yards of cloth.

STATEMENT V—(*continued*).

Exports of Raw Jute and Manufactures to Foreign Countries, during 1929-30.

Countries.	Raw Jute. Tons.	Twist and Yarn in thousand lbs.	Gunny Bags (*a*) in thousand Nos.	Cloth in thousand Yards.
Indo-China ...	......	12,029	115	......
Iraq	......	1,577	523	......
Italy	54,902	......	......	......
Japan	16,007	15,742	120	......
Java	......	37,289	850	22
Kenya Colony ...	......	5,507	1,189	......
Madagascar ...	......	2,161	......	......
Malay	......	213	......	......
Mauritius	......	3,677	......	......
Mexico	2,146	165	......	......
Morocco	......	1,574	......	......
Natal	......	15,126	4,533	191
Netherlands ...	21,630	1,610	282	......
New Zealand ...	......	10,949	3,829	......
Nicaragua	......	408	......	......
Norway	......	807	2,402	......
Panama	......	102	......	......
Paraguay	......	......	571	......
Persia	......	417	114	......
Peru	......	9,216	2,539	118
Philippines	......	7,109	14,591	......
Porto Rico ...	......	......	210	......
Roumania	......	1,078	......	......
Russia	497	838	......	......
Sansah-badra ...	......	285	......	......
Serbcrot	......	330	......	......
Siam	......	10,510	......	......
Smyrna	......	927	876	......
Somaliland ...	......	178	......	......
Spain	44,720	......	......	......
Straits Settlements ...	......	13,877	1,329	117
Sudan	......	4,358	......	......
Sumatra	......	832	827	......
Sweden	......	......	170	......
Syria	......	164	......	......

(*a*) One gunny bag represents 2 yards of cloth.

STATEMENT V—(*concluded*).

Exports of Raw Jute and Manufactures to Foreign Countries during 1929-30.

Countries.	Raw Jute. Tons.	Twist and Yarn in thousand lbs.	Gunny Bags (*a*) in thousand Nos.	Cloth in thousand Yards.
Tanganyika ...	......	627	336	......
Turkey	......	1,953	2,853	......
United Kingdom ...	164,751	53,046	58,593	298
United States ...	79,386	16,972	1,071,909	......
Uruguay	......	......	14,612	166
Venezuela	......	......	100	......
West Indies ...	......	3,513	......	......
Zanzibar	......	397	107	......
TOTAL (including Exports to less important countries) during 1929-30	806,884	522,291	1,650,526	6,009
TOTAL of the year 1930-31	619,705	434,046	1,271,000	7,341

(*a*) One gunny bag represents 2 yards of cloth.

STATEMENT VI.

Indian Consumption of Jute Fabrics and Yarn.

Figures in Thousands.

Year.	Gunny Bags.		Cloths.	
	Nos.	Value in Rupees.	No of Yards.	Value in Rupees.
1931	58,173	14,345	26,182	2,922
1930	62,298	19,126	24,333	3,410
1929	61,883	23,447	28,795	5,396
1928	71,135	29,140	43,594	8,109
1927	79,000	33,659	59,675	9,915
1926	79,522	41,144	54,900	10,556
1925	81,658	43,747	54,814	10,839
1924	94,951	41,905	43,796	7,627
1923	83,271	35,164	39,228	7,138
1922	57,760	21,491	13,520	2,604
1921	70,560	27,226	18,098	3,134
1920	103,807	48,294	32,613	7,641

STATEMENT VII.

Total Value of Exports of Raw Jute and Manufactures from India.

Year.	Value of Jute.	Value of Manufactures.	Total Value of Exports.
	Rupees in thousands.	Rupees in thousands.	Rupees in thousands.
1931-32 ...	11,18,81	21,92,42	33,11,23
1930-31 ...	12,88,47	31,89,44	44,77,91
1929-30 ...	27,17,38	51,92,68	79,10,06
1928-29 ...	32,34,92	56,90,49	89,25,41
1927-28 ...	30,66,26	53,56,43	84,22,69
1926-27 ...	26,78,04	53,18,09	79,96,13
1925-26 ...	37,94,57	58,83,98	96,78,55
1924-25 ...	29,09,30	51,76,66	80,85,96
1923-24 ...	20,00,06	42,28,36	62,28,42
1922-23 ...	22,52,85	40,49,42	63,02,27
1921-22 ...	14,04,92	29,99,57	44,04,49
1920-21 ...	16,36,09	52,99,47	69,35,56
1919-20 ...	24,69,95	50,01,55	74,71,50
War Period—1914 to 1918.			
1918-19 ...	12,72,01	52,65,23	65,37,24
1917-18 ...	6,45,38	42,84,31	49,29,69
1916-17 ...	16,28,81	41,67,24	57,96,05
1915-16 ...	15,64,20	37,97,85	53,62,05
1914-15 ...	12,91,02	25,82,03	38,73,05
Pre-war Period—			
1913-14 ...	30,82,64	28,27,37	59,10,01
1910-11 ...	15,49,00	16,99,49	32,48,49
1901-02 ...	11,79,73	8,71,14	20,50,87

STATEMENT VIII.

Value of Jute and Jute Manufactures compared with other Principal Products of the Trade.

Year.	Value of the Principal Indian Products exported.	
	Rupees in Thousands.	
1930-31 ...	2,204,553	Of the total fall-off of Rs. 903,502,000 in value of the Exports of Indian Products, during 1930-31, Jute and Jute Manufactures were affected to the extent of Rs. 411,230,500, in comparison with the average annual value of the exports of jute fibre and its products of the previous five years.
1929-30 ...	3,108,055	
1928-29 ...	3,301,279	
1927-28 ...	3,191,535	
1926-27 ...	3,014,358	
1925-26 ...	3,748,421	

STATEMENT IX.

Comparative Values of Total Exports of the Principal Indian Products.

Indian Merchandise.	Percentage of the Value of Total Exports. 1929-30.	1928-29.	1927-28.
Jute, Raw	8·74	9·80	9·61
Manufactures	16·71	17·24	16·78
Both ...	25·45	27·04	26·39
Cotton, Raw	21·11	20·20	15·09
Manufactures	2·31	2·36	2·72
Both ...	23·42	22·56	17·81
Food Grains*	11·90	10·21	13·45
Seeds	8·52	8·97	8·36
Tea	8·37	8·06	10·18
Metals and Ores	3·33	2·90	2·81
Leather	2·63	2·82	2·84
Hides	2·57	2·70	2·76
Lac	2·54	2·62	2·19
Wool	1·72	1·79	1·67
Wax	1·02	1·16	0·76
Oil Cake	1·00	0·74	0·98
Other Minor Merchandise below 1% each	7·53	8·43	9·80

* Food grains consist of Rice, Wheat, Pulses, Barley, Jowar and Bajra, of which Rice has a share of about 10% in the total value.

STATEMENT X.

Number of Jute Mills, Looms, Spindles and Labourers employed.

Year.	Jute Mills. No.	Looms. Hessian.	Looms. Sacking.	Total Looms.	Spindles. In Thousands.	Persons employed. Daily Average Nos. in Thousands.
BENGAL.						
1930 ...	89	36,245	21,894	58,639*	......	......
1929 ...	87	31,534	19,502	51,036	10,85	3,37
1928 ...	87	31,184	19,337	50,521	10,84	3,29
1927 ...	87	31,001	19,353	50,354	10,61	3,27
1926 ...	85	31,307	18,541	49,848	10,49	3,27
1925 ...	85	31,055	18,344	49,399	10,54	3,37
1924 ...	84	29,732	18,286	48,018	10,30	3,26
1923 ...	77	28,111	17,452	45,563	......	......
1922 ...	77	24,985	16,354	41,339	8,97	2,85
1921 ...	72	24,453	16,445	40,898	8,63	......
1920 ...	72	24,353	16,124	40,477	8,63	......

Of the total Calcutta looms, 62% being roughly hessian and 38% sacking. During the year ended with 30th June, 1930, when the Calcutta mills worked generally 60 hours a week, each loom consumed 532·5 maunds of raw jute. On a daily average, each loom employs 6·5 persons, but attempts have been successful in reducing the number to 5.

MADRAS.						
1930 ...	...	......	......	......	......	......
1929 ...	4	......	......	889	20·6	5·9
1928 ...	4	......	......	807	17·2	5·6
1927 ...	4	......	......	753	17·6	5·5
1926 ...	4	......	......	566	13·2	4·4
1925 ...	4	......	......	565	12·9	4·5
1924 ...	4	......	......	565	12·9	4·6
1922 ...	3	......	......	496	11·5	3·7
1921 ...	3	......	......	296	6·5	......

The first mill was established in Madras in 1901.

BIHAR & ORISSA.						Actual Average.
From 1925 to 1930	1	......	......	100	2·5	842

UNITED PROVINCES.

There was one private mill with 75 looms which commenced work in 1923 and closed down in 1928.

*Excepting about 2,000 looms possessed by the Agarpara Mill and the three Ludlow mills.

STATEMENT XI.

Prices of Raw Jute in Calcutta.

(*Season July to June.*)

Year.	Loose Jute. 4's—District Quality. Per maund (80 lbs.)			Baled Jute. Firsts–(*M*). Per bale of 5 maunds (400 lbs.).		
	July.	October.	January.	July.	October.	January.
	Rs. A.	Rs. A.	Rs. A.	Rs. A.	Rs. A.	Rs. A.
1930-31 ...	7 8	5 12	6 0	45 8	30 0	28 4
1929-30 ...	11 0	12 0	10 12	62 0	60 0	58 8
1928-29 ...	14 0	11 8	13 0	74 0	67 0	66 0
1927-28 ...	11 0	11 4	12 0	61 8	63 0	64 0
1926-27 ...	13 8	12 0	......	82 0	58 0	62 8
1925-26 ...	15 12	23 0	25 0	95 0	112 0	123 0
1924-25 ...	12 0	15 4	16 0	61 8	75 8	83 0
1923-24 ...	10 4	8 0	11 4	61 0	45 0	63 0
1922-23 ...	17 12	14 8	16 0	88 0	83 0	86 0
1921-22 ...	11 0	13 4	12 4	61 0	69 0	65 0
1920-21 ...	10 4	13 0	12 8	65 0	86 0	74 0
1919-20 ...	......	......	......	81 0	......	......
War Period 1914-15 to 1918-19.						
1918-19 ...	......	......	......	43 0	......	77 0
1917-18 ...	......	......	......	40 0	......	37 0
1916-17 ...	......	......	......	48 8	......	53 0
1915-16 ...	......	......	......	48 0	......	48 0
1914-15 ...	......	......	......	67 0	......	42 0
Pre-war Year 1913-14.						
1913-14 ...	......	......	......	69 0	83 0	85 8
1910-11 ...	......	......	......	33 0	......	50 0
1900-01 ...	......	......	......	35 0	......	30 0

STATEMENT XII.

Wholesale Prices of Manufactured Goods in Calcutta

(Season July to June.)

Year.	Hessian, $10\frac{1}{2}$ oz., 40", per 100 yards.			Gunny Bags (Twill), $2\frac{1}{4}$ lbs., $44'' \times 26\frac{1}{2}''$, per 100.		
	July.	October.	January.	July.	October	January.
	Rs. A.	Rs. A.	Rs. A.	Rs. A.	Rs. A.	Rs. A.
1930-31 ...	13 10	10 12	10 2	33 8	29 0	27 4
1929-30 ...	17 12	17 10	13 14	34 8	36 4	31 8
1928-29 ...	24 14	21 6	20 12	47 8	45 4	46 12
1927-28 ...	20 15	22 2	23 14	44 12	47 0	47 4
1926-27 ...	20 2	18 14	19 12	48 0	51 4	49 0
1925-26 ...	23 2	26 6	27 0	57 0	71 4	69 8
1924-25 ...	21 10	24 4	22 2	52 12	65 8	63 8
1923-24 ...	18 4	18 4	18 8	45 8	44 0	48 4
1922-23 ...	28 12	24 0	23 8	46 8	45 0	47 8
1921-22 ...	13 12	17 4	15 4	31 12	42 0	39 4
1920-21 ...	24 8	26 8	16 8	41 8	45 8	34 0
1919-20 ...	28 0	......	......	61 0	......	......
War Period 1914-15 to 1918-19.						
1918-19 ...	28 0	......	21 0	67 0	......	58 0
1917-18 ...	45 0	......	49 0	43 0	......	68 0
1916-17 ...	18 0	......	17 8	40 8	......	44 8
1915-16 ...	17 4	......	19 4	43 8	......	43 0
1914-15 ...	22 0	......	11 0	42 8	......	34 0
Pre-war Year 1913-14.						
1913-14 ...	17 0	21 10	17 0	33 8	41 0	37 4
1910-11 ...	9 3	......	10 0	27 4	......	30 0
1900-01 ...	10 6	......	11 0	22 0	......	23 0
1873 ...	March. 11 0	......	......	March. 22 0	......	......

STATEMENT XIII.

Area under Principal Crops in British India Cultivated during 1929-30

Net Area Sown ... ... 228,160,000 acres.

Current Fallows ... ... 49,714,000 acres.

Total Acreage under Crops including double-cropped area 260,344,000

Crop.	Area in acres. In Thousands.	Percentage.	Crop.	Area in acres. In Thousands.	Percentage.
Food Crops :—			**Oil Seeds :—**		
Rice ...	79,424	30·5	Linseed ...	1,927	0·7
Wheat ...	24,731	9·5	Sesamum ...	3,556	1·4
Barley ...	7,027	2·7	Rape and		
Jowar ...	23,241	8·9	Mustard ...	3,555	1·4
Bajra ...	13,291	5·1	Groundnut ...	5,012	1·9
Ragi or Marua	4,000	1·5	Cocoanut ...	654	0·3
Maize ...	6,552	2·5	Castor ...	436	0·2
Gram ...	11,458	4·4	Other Oil Seeds	1,190	0·5
Other food grains	30,294	11·6			
Sugar ...	2,583	1·0	TOTAL OIL SEEDS	16,330	6·3
Miscellaneous Food Crops ...	7,898	3·0			
TOTAL FOOD CROPS	210,499	80·8	**Drugs and Dyes :—**		
			Tea ...	766	0·3
Fibres :—			Coffee ...	91	0·04
Cotton ...	16,141	6·2	Indigo ...	71	0·03
Jute ...	3,268	1·2	Opium ...	41	0·02
Other Fibres			Tobacco ...	1,172	0·5
(San and Deccan hemps)...	666	0·3	TOTAL ...	2,141	0·9
			Fodder Crops	9,381	3·6
TOTAL FIBRES ...	20,075	7·7	Miscellaneous Non-food Crops	1,918	0·7
			Total Non-food Crops ...	49,845	19·2

The Total Area sown in the Indian States, 66,520,000 acres, of which under—

Wheat ...	6,000,000	Opium ...	21,000
Rice ...	3,850,000	Coffee ...	114,000
Cotton ...	7,285,000	Tea ...	64,000
Other Fibres	163,000	Tobacco ...	209,000

STATEMENT XIV.

Railway and Steamer Tariffs.

Distance. Miles.	Stations.	Description of articles. (a)	Rate per bale. Busy season (July to December).	Rate per bale. Slack season (January to June).
	By Railway and Steamer combined :—			
271	Mymensingh to Calcutta (viâ Jagannathganj and Serajganj by Str.)	3½ mds. bale	Str. 0 4 11 Ry. 1 12 3	0 3 6 1 6 0
	Ditto	1½ mds. bale	Str. 0 2 1 Ry. 0 12 1	0 1 6 0 9 4
322 (b)	Dacca to Calcutta (viâ Narayanganj and Khulna by Str.)	3½ mds. bale	Str. 0 15 5 Ry. 0 12 10	0 11 0 0 9 4
	Ditto	1½ mds. bale	Str. 0 6 7 Ry. 0 5 6	0 4 8 0 4 0
	By Railway only :—			
197	Serajganj to Calcutta	3½ mds. bale	1 12 11	1 4 4
	Ditto	1½ mds. bale	0 12 5	0 8 8
	By Railway, all the year round :—			
271	Rangpur to Calcutta	3½ mds. bale	2 6 3	......
	Ditto	1½ mds. bale	1 0 5	......
348	Kissenganj to Calcutta	3½ mds. bale	2 9 11	......
	Ditto	1½ mds. bale	1 1 11	......
155	Goalundo to Calcutta	3½ mds. bale	1 4 4	......
	Ditto	1½ mds. bale	0 8 8	......
109	Khulna to Calcutta	3½ mds. bale	0 15 10	......
	Ditto	1½ mds. bale	0 6 10	......
	By Steamer (I. G. S. N. Co. and R. S. N. Co.) :—			
492	Narayanganj to Calcutta (Nimtolaghat)	4 mds. bale	2 2 1	1 8 0
	Ditto	3½ mds. bale	1 14 10	1 5 8
	Ditto	1½ mds. bale	0 13 2	0 9 4
544	Serajganj to Calcutta (Nimtolaghat)	Same rates as	above.	
455	Chandpur Do.	Do.	do.	
415	Madaripur Do.	Do.	Do.	

(a) A bale of 4 maunds or 3½ maunds is to be pressed to 13·5 c. ft. or less, and that of 1½ maunds pressed to 5·88 c. ft. or less.

(b) Distance from Dacca to Calcutta, *via* Narayanganj and Goalundo, is 270 miles.

STATEMENT XV.

SHIPPING FREIGHT SCHEDULE.

Rates per Ton from Calcutta : usually a Concession of 10% allowed from January to July.

Ports.	Raw Jute.	Gunny.	Ports.	Raw Jute.	Gunny.
	s. d.	*s. d.*		Rs. A.	Rs. A.
London ...	31 3	35 0	Kobe ...	30 0	30 0
Liverpool ...	31 3	35 0	Bankok ...	32 0	30 0
Dundee ...	31 3	35 0	Hankow ...	......	46 6
Glasgow ...	31 3	35 0	Vladivostok	......	38 0
Marseilles ...	31 3	35 0	Fiji ...	......	130 0
Dunkirk ...	30 0	32 6			
Antwerp ...	27 6	30 0		*s. d.*	*s. d.*
Hamburg ...	27 6	30 0			
Bremen ...	27 6	30 0	Sydney ...	37 6	50 0
Italy ...	27 6	37 6	Melbourne ...	37 6	50 0
New York ...	32 6	35 0	New Zealand	......	70 0
Boston ...	32 6	35 0	Port Elizabeth	......	52 6
Philadelphia	32 6	37 0	Cape Town	......	55 0
Santos ...	60 0	......	Zanzibar ...	......	48 12
Chile ...	......	40 0	Mozambique	......	91 6
Peru ...	......	65 0	Tanga ...	......	70 0
San Francisco	......	......			
				Rs. A.	Rs. A.
	Rs. A.	Rs. A.	Mombassa		
			(*via* Bombay)	......	48 12
Manila ...	......	36 8	Mauritius ...	......	32 0
Havana ...	......	55 0			
Jamaica ...	......	70 0			*s. d.*
Barbadoes ...	......	80 0			
Porto Rico	......	80 0	Madagascar	......	95 0
Colombo ...	......	15 0	Basra ...	......	32 8
Karachi ...	......	20 0	Muscat ...	......	32 8
Bombay ...	......	16 0	Soudan ...	......	26 0
Madras ...	......	16 0	Port Said ...	......	15 0
Rangoon ...	......	14 0	Suez ...	......	22 6
Java ...	......	15 0	Alexandria ...	......	15 0
Penang ...	20 0	20 0	Constanti-		
Singapore ...	20 0	20 0	nople ...	......	15 0
Hongkong ...	25 0	25 0	Odessa ...	......	26 0
Shanghai. ...	30 0	30 0	Smyrna ...	......	21 0

APPENDIX A.

Names of the Jute Mills working in Bengal during 1930.

Mills.	Looms.	Agents.
Adamjee Jute Mills Co., Ld.	652	Adamjee Hajee Dawood & Co.
Albion Jute Mills Co., Ld. ...	498	Andrew Yule & Co., Ld.
Alexandra Jute Mills Co., Ld.	423	Begg, Dunlop & Co., Ld.
Alliance Jute Mills Co., Ld. ...	1,232	Do.
Anglo-India Jute Mills Co., Ld.	2,561	Duncan Bros. & Co., Ld.
Angus Jute Works ...	1,196	The Angus Co., Ld,
Auckland Jute Co., Ld. ...	810	Bird & Co.
Balliaghata Jute Mills ...	205	James Luke & Sons.
Bally Jute Co., Ld. ...	825	George Henderson & Co.,Ld.
Baranagore Jute Factory, Ld.	1,744	Do.
Belvedere Jute Mills, Ld. ...	820	Andrew Yule & Co., Ld.
Birla Jute Manufacturing Co., Ld.	1,301	Birla Bros., Ld.
Budge-Budge Jute Mills Co., Ld.	990	Andrew Yule & Co., Ld.
Caledonian Jute Mills Co., Ld.	616	Do.
Champdany Jute Co., Ld. ...	544	James Finlay & Co., Ld.
Do. (Wellington Mill) Branch ...	796	Do.
Cheviot Mills Co., Ld. ...	530	Andrew Yule & Co., Ld.
Clive Mills Co., Ld. ...	969	Bird & Co.
Craig Jute Mills Ld. ...	368	Begg, Dunlop & Co., Ld.
Dalhousie Jute Co., Ld. ...	766	Bird & Co.
Delta Jute Mills Co. ...	799	Andrew Yule & Co., Ld.
Empire Jute Co., Ld. ...	500	McLeod & Co.
Fort Gloster Jute Manufacturing Co., Ld.	1,884	Kettlewell Bullen & Co., Ld.
Fort William Jute Co., Ld....	1,014	Do.
Gagalbhai Jute Mills, Ld. ...	600	Mafatlal Gagalbhai
Ganges Manufacturing Co., Ld.	2,309	Macneil & Co.
Gourepore Co., Ld. ...	1,392	Barry & Co.
Hastings Mills	1,077	Birkmyre Bros.
Hoogly Mills Co., Ld. ...	454	Gillanders Arbuthnot & Co.
Howrah Mills Co., Ld. ...	1,940	Jardine Skinner & Co.
Hukumchand Jute Mills, Ld.	1,751	Sir Sarupchand Hukumchand & Co.
India Jute Co., Ld. ...	1,076	Mackinnon, Mackenzie & Co.
Kamarhatty Co., Ld. ...	1,797	Jardine Skinner & Co.
Kanknarrah Co., Ld. ...	1,527	Do.
Kelvin Jute Co., Ld. ...	702	McLeod & Co.
Khardah Co., Ld. ...	1,457	Anderson Wright & Co.
Kinnison Jute Mills Co., Ld.	1,250	F. W. Heilgers & Co.
Lansdowne Jute Co., Ld. ...	846	Bird & Co.
Lawrence Jute Co. ...	840	Do.
Lothian Jute Mills Co., Ld. ...	558	Andrew Yule & Co., Ld.

Mills.	Looms.	Agents.
Megna Mills Co., Ld. ...	1,008	Mackinnon, Mackenzie & Co.
Naihati Jute Mills Co., Ld. ...	815	F. W. Heilgers & Co.
National Co., Ld. ...	852	Andrew Yule & Co., Ld.
New Central Jute Mills Co., Ld.	709	Do.
Northbrook Jute Co., Ld. ...	567	Bird & Co.
Nuddea Mills Co., Ld. ...	1,092	Barry & Co.
Orient Jute Mills Co., Ld. ...	601	Andrew Yule & Co., Ld.
Premchand Jute Mills, Ld. ...	500	Raja Janakinath Roy & Bros.
Presidency Jute Mills Co., Ld.	395	McLeod & Co.
Reliance Jute Mills Co., Ld. ..	1,276	Jardine Skinner & Co.
Samnaggar Jute Factory Co., Ld.	2,013	Thomas Duff & Co., Ld.
Shree Hanuman Jute Mills, Ld.	660	Sooraimull Nagarmull
Societe Generale Industrielle de Chandernagor (Gondalpara)	600	Gillanders Arbuthnot & Co.
Soorah Jute Mills Co., Ld. ...	395	McLeod & Co.
Standard Jute Co., Ld. ...	650	Bird & Co.
Titagarh Jute Factory Co., Ld.	1,794	Thomas Duff & Co., Ld.
Union Jute Co., Ld. ...	1,326	Bird & Co.
Victoria Jute Co., Ld. ...	1,331	Thomas Duff & Co., Ld.
Waverley Jute Mill Co., Ld. ...	456	Begg Dunlop & Co., Ld.

Altogether, there were 89 mills with 58,639 looms and 1,085,000 spindles at work, in Bengal, during the year 1929-30, under 59 registered concerns, with the exception of 3 Ludlow mills of Ludlow Jute Co., Ld., Chengail, District Howrah, and the Agarpara Jute Mill (Messrs. B. N. Elias & Co.), who are not members of the Indian Jute Mills Association; while, during 1917-18, there were 72 mills, 40,271 looms and 825,795 spindles.

Names of the Jute Mills of the other Provinces in India.

Names of the Mills.	Address.
MADRAS.	
Chittivalsah Jute Mills	Vizagapatam.
Nellimarla Jute Mills	Vizagapatam.
Shri Krishna Jute and Cotton Mills ...	Ellore, West Godavari.
Guntur Jute, Cotton and Paper Mills Co., Ld.	Guntur.

There are 4 mills in Madras with 889 looms and 20,600 spindles employing about 6·5 persons per loom per diem.

BIHAR.	
Rameswar Mills	Muktapur, Darbhanga.

The mills are equipped with 100 looms and 2,500 spindles employing 312 labourers per diem.

APPENDIX B.

Shippers of Raw Jute and Jute Manufactures, Calcutta.

Adamjee Hajee Dawood & Co., Ld.
Anderson Wright & Co.
Andrew Yule & Co., Ld.
Anglo-Oriental Bag Co.
Arnold & Co., Ld.
B. C. Choraria
Badri Dass Fulchand
Becker Gray & Co., Ld.
Blackwood Blackwood & Co.
Bird & Co.
B. Meyer & Co., Ld.
B. N. Elias & Co., Ld.
Birkmyre Brothers
Birla Brothers, Ld.
Chhajuram & Sons.
D. L. Miller & Co.
David Sassoon & Co., Ld.
Earl, Acton & Co., Ld.
E. Meyer & Co., Ld.
Gangjee Sajun & Co.
George Henderson & Co.
Gladstone, Wyllie & Co.
G. & M. Fogt
G. M. R. L. Gouti
Graham Trading Co.
H. B. Sodhani & Co.
Hoare Miller & Co., Ld.
H. N. Chand
Imperial C. Industries, Ld.
James Gow & Co.
James Finlay & Co., Ld.
James Scott & Sons, Ld.
Japan Cotton Trading Co.
Jardine Skinner & Co.
Jaydoyal Madan
Jiwanram Johar Mull
J. C. Duffus & Co., Ld.
J. H. Zeissen & Co., Ld.
J. M. C. Mull
Jute Industries, Ld.
Kain Gwan & Co., Ld.
Kettlewell, Bullen & Co., Ld.
Louis Dreyfus & Co.
Mafatlal Gagalbhai & Sons
M. A. Sassoon & Sons, Ld.
McLeod & Co.
Mahilaram Ramjee Das
Maluchand Surajmull
Mitsui Bussan Kaisha, Ld.
M. M. Ispahani & Sons
Ralli Brothers, Ld.
Rammal Lillosha
R. D. R. K. Das
R. Steel & Co., Ld.
S. C. Chander
B. D. Ramjee Dass
Senda & Co., Ld.
Shaw Wallace & Co., Ld.
Soorajmull Nagarmull
S. Manasseh & Sons
The Angus Co., Ld.
The Bombay Co., Ld.
The Calcutta Steam Navigation Co., Ld.
The Chittagong Co.
The Ludlow Jute Co., Ld.
The Narayangung Co.
Turner Morrison & Co., Ld.

INDEX.